# The Evolution of Sex

*A New Approach*

Using Both Pathways of Evolution

Sean Gould

The Evolution of Sex: A New Approach



Amazon Publishing
• USA 2019

**ISBN:** 9781090183033
**Imprint:** Independently published



This printing; Aug 2019
8/2019 1Sex0

Other books by Sean Gould (see note below);

*Maximizing Options*
*How Evolution Can Explain Human Behavior*

"The reasoning here is persuasive and his knowledge of the background is very impressive and it all results in a theory of human evolution and behavior that is in some ways very unique and this is a splendid contribution to the human evolution literature."

Todd Stark

(Todd is a contributor to the evolution debate and a top 500 reviewer on Amazon.)

"This volume provides a valuable discursive evolutionary explanation for the emergence of moral and ethical reasoning - a topic little, if ever, offered much space in the modern literature concerned with evolutionary psychology and adaptive intelligent systems."

Tony Dickinson

(Tony is a visiting research fellow to the Snyder Lab of the McDonnell Center for Higher Brain Function at Washington University School of Medicine.)

"There are aspects of evolutionary theory that do not yet explain certain human behaviors all that well. The ideas expressed in the book are timely, and will increase our knowledge of the human condition, and the processes that operate in the universe; something in which we all seem so interested. I was impressed."

Roger McEvilly

(Roger is a contributor to the evolution debate, and is a top 1,000 reviewer on Amazon.)

Note: This book was published as *The Theory of Options; A New Theory of the Evolution of Human Behavior*. It is intended to change the title to *Maximizing Options,* and some parts of the book will be rewritten.

This book says that the model in evolution is incomplete. The criticism at first seems simple.

Evolution works by natural selection. Models of molecular and cellular selection prove how it works. Except, if you study the theory, not just any book, but all of them, there is no model of molecular selection that fully works. It is not just to formulate the theory. Rather, the model of evolution must be changed to prove the new workings.

This book provides the new model.

Anyone can criticize, test the theory, or consult other experts. The test concerns, is it correct, can it explain sex, and the other paradoxes of the theory?

## Table of Contents

## ***Controversies ...***

This book offers new model of evolution, enhanced over the standard. To most people, life is via Darwinism. Genes compete for fitness, and the advanced genes spread across life.

In my theory, non-life also began by natural selection. This was about 4.1 to 3.8 billion years ago. However, full Darwinism did not start until 3.7 to 3.5 billion years. This idea of non-life by first selection, leading then to Darwinism is not new. It is like the standard, but with a slight variation in the words.

Except, now I have an extraordinary suggestion.

Please check with experts, not in any book, but all of them. Darwinism was always along a real axis. However, I propose that the first selection was imaginary. Now, this will affect the math, but having first selection as imaginary can solve several problems.

Now, to an expert this does not seem credible. Most people are not even aware the selection can be both real and imaginary. To them, Darwinism has been proven, and selection is real. Darwinism, which requires fitness, is not in early selection. It was added in a pre-Darwinian stage. However, a more modern instance, three quarters of the way to modern life sex had evolved. Now, it seems simple, but sex has not been fully explained, so how do you explain that?

Well, sex started not at 4 billion years ago, but three quarters of the way through. By 2.2 to 1.2 billion years organisms reproduced directly, except there was no sex. At 1.2 billion years sex occurred, from meiosis adding onto mitosis. It will be explained, but two organisms now had to form a pair to reproduce. This 2 for 1 reproduction seemed to cause a 50% loss of genes passed on. Moreover, for a billion years there was no great increase in genome size, but with sex genome size increased. There is no answer for an increase in the genome size.

Especially, the standard theory cannot explain sex. The genome size has increased slightly pre-sex, but with sex, there is a huge increase. Experts have tried to explain this, but with no result. There must be another effect on gene mutability, because the mutation rate is known, but it is not used in an equation. If you assume that the spread of genes is always real, there cannot be where one part is real, but the other imaginary. However, if just one part is real, the imaginary effects can combine.

Still, let me caution on the role of genes. If a gene began in prelife, it follows, I will call it, a basic math. If a gene started to spread, from prelife, you can trace its spread, and can write the equation. This is how I solved for the prelife spread of early genes. Richard Dawkins wrote that the first genes would be basic, but he never developed it. The math was there, but he could not see how it would continue.

Instead, experts used an advanced effect to imply the results, but I used a basic math to show the spread. This is how I obtained the two axis of gene spread. An advanced math will not prove the sudden increase in gene spread, once sex began, because the gene increase is by an act of imaginary spread for the second axis. For example, the histone H4 is on an imaginary axis, and has a rate of 98% of normal genes. Shorter genes keep higher rates to the start of life. Using this, I will show how the spread of genes resulted in a logistic curve. Here, the rate of gene spread is already known, but without an effect of fitness.

For me, all other attempts apart from this to explain sex fail. It does not matter how many computers are used or who is an expert, without breaking it into two axes of gene spread, the calculation will not add up. If you can settle its workings, sex can advance. Other problems of the gene dispersion could also be solved.

Well, this is my suggestion, but could I be wrong?

Evolution is a topic for which it is easy to be wrong. Evolution is both a fact and a theory. The facts can be checked, at a minimum, on Wikipedia or other sources. But evolution is also a theory. The theory can be checked on a computer. I ran mine in basic math, but a more advanced program can also be used to check the results.

The problem again is that there is no molecular selection, proven in an equation. You can check this in any presentation. My model shows how this causes many paradoxes, so even if my model is proven incorrect, it does not solve the paradoxes that are entrenched. My model can solve this by selection along a real and imaginary axis.

If this is wrong my thesis has failed, though check it in many areas.

## *The Evolution of Sex ...*

This book has a new model for the evolution of sex.

The thesis is that evolution, as now formed, is incomplete. It has a well-tested model of *cellular* selection, but there is no corresponding model of *molecular* selection. In the theory, cellular selection had caused molecular selection, but this is not as per facts. Molecular selection was primary, or first, and this resulted in cellular selection.

For example, life evolved from 4.1 to 3.8 billion years. Then, cellular selection was from near 3.7 to 3.5 billion years. Yet what occurred from about 4 until 3.5 billion years. It was by natural selection, and experts can right the equation, but there is no model from early days forward. Experts state that this was in the past, and we do not know all the details. Except, the problem is the equation itself. Unless you can write the full equation, the rest of evolution will not work.

Yet, I can write the equation.

It is by natural selection. It lines up, very well, with the other facts of how life evolved. The secret is to split the two into the real axis and an imaginary axis. For instance, RNA, DNA, ATP, ADP, and the ribosome evolved very early. They hold a similar structure across early life and were also formed before cellular selection began. The secret is to allow these to evolve along an imaginary axis. This can explain, by natural selection, how early molecules formed. It can also explain how the first molecules came, by natural selection, along the extra axis. This, and several effects, can now explain how evolution works.

As well, my theory can explain sex.

Now, evolution works as it is, with real numbers alone, with no need for an imaginary component. Except, how do you explain sex? When sex evolved there was loss, from 100% of genes passed down to 50%, by a reduction of 2 to 1 of genes. With sex the copying rate increase though, from a narrow line to a wide increase of types. How do you explain this? If you are an expert, if sex falls from 100% of genes passed on to 50%, why does sex go up? If you have written books on sex, you cannot explain this. If you are a professor, you cannot explain it. If you studied the math, you still cannot explain it.

Except that, using my theory, it will work. If you do not know how sex works, look at the next diagram.

It is in Excel. Perhaps experts do not follow, but the front black curve is how genes manipulate on one axis. However, the crux is the back curve. It shows a far grey imaginary line of how genes try to spread in many populations. The problem in sex is to move from the front black curve to the far grey curve. The key is the mutation rate, which is added to indicate the rate of mutations in gene spread. Of course, from sex as a theory, to how it evolved is difficult. If we develop the back line, it will fall as gene mutation increases. If genes from the front spread for many populations, the cut-off is a rate about $6 \times 10^{-9}$ mutations. Incredibly, as the back-line falls, genes seem to maintain this high rate of mutation Especially, this is for the early curves, which retain the mutation rate in many species.

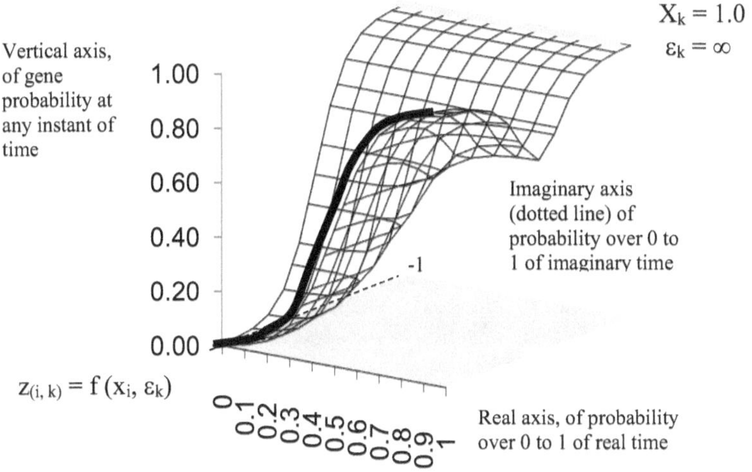


Fig 0.1 This shows a gene gaining along a single left vertical axis, but the axis is unfolded along two pathways of time. The front axis shows 'real' gene spread. The rear axis shows spread along an 'imaginary' time, across many populations.

If you read a book on evolution, it calls for a combination. What I call, *cellular* selection must add onto *molecular* selection. It is in each book, but if you study the books, there is no molecular selection. Still, there is molecular selection and its workings, and the *neutral* theory, so it seems absurd to claim that it is not there.

Nevertheless, there is proof of a mistake.

Four billion years ago, before true cells evolved, before they split to archaea and bacteria, natural selection had to start. Assume that it started with molecular selection. This sorted which molecules could evolve first. Now, unlike in other books, my theory of molecular selection works. It uses a logistic curve, and it runs from 0% to 100% of gain. You can test this. When cells began about 3.5 billion years ago, then molecular and cellular selection had to add together. The cellular selection also uses a logistic curve, and it runs from 0% to 100%.

Now, I can claim that there is no model of molecular selection. I can show a model of how this worked, and any person can test it. I can state that molecular and cellular selection adds together, as per the books. Now, to keep the math simple, I made them run from 0% to 100% along each axis. However, you cannot state that selection began four billion years ago, but it did not start until 3.5 billion years ago. What happened from 4 to 3.5 billion years? You can claim it was by natural selection, but no one has the equation. You must choose right at the start, which equation that you will use for the rest of evolution.

Still, how do I prove that this new model can work?

I require two things. First, I need a table, of the most conserved genes down to the most variable, across all life. I have parts of table, but a full table is not drawn. There is a billion-fold difference ($10^{-4}$ to $10^{-13}$) between the most conserved and the variable genes.

Then, I need an equation, of how the two parts add. Now, experts will insist there cannot be such an equation. One equation, in genes, cannot show how part of evolution spreads by conserving sequence, but another part shows how genes compete as species. However, it only applies for selection along one axis, but I use two axes of time. My equation shows the ratio of conserved to variable parts of genes. It reproduces the curve in Fig 0.1, and this is not in other texts. The molecular selection applies over life. By contrast, cellular selection is for species, such as a mouse can be better adapted than a giraffe.

Except, what proof do I have of this 'real' and 'imaginary' axis. Well, it is a fact. Experts can object I should not use this, but the split of genes into real and imaginary is correct. Anyone can test it. Real and imaginary are in minor effects, but no one has gone to the next phase, to consider that it could apply across life. Right at the start, billions of year ago it split into real and imaginary, and this applied to life. Anyone who doubts this can try their own equations, but mine works.

Still, an expert might insist that the curve cannot work. The back curve is that the higher gene conservation, the more the gene is conserved (the most conserved gene is the longest to exist). To a modern species, the higher mutation seems to detract from conservation. Rather, the back curve is how conserved genes since first life, the higher it is conserved. Yet to do this, I did not study mathematics, or evolution, but I did study electrical engineering. The cellular selection is along a real axis, but the molecular selection is for an imaginary axis. In the universe, facts are the expansion across galaxies. In electrical engineering, facts are that it is different for alternating and direct current. In evolution, facts concern genes. Again, the universe is not my specialty, yet one must add 100% plus 100%, but the two cannot exceed 100%.

The answer to sex, then is in the mutation rate. If the gene has a high rate, above $10^{-13}$, it spreads across higher life, such as in eukarya. If it has a medium rate, about $6 \times 10^{-9}$, it also allows sex to spread in higher life. If the rate is below $6 \times 10^{-9}$, the curve will revert to a lower value. Roughly, the standard model of evolution works, but only for the mutation less that the $6 \times 10^{-9}$. For the $6 \times 10^{-9}$ rate, for first life, and the evolution of sex, you must use my new theory.

When Darwin explained it, there was no math, so he gave a verbal model of evolution. Later, genes and the math were discovered, but there were controversies, so the emphasis was to prove that the theory backed the description by Darwin. People saw controversy, such as the evolution of sex, the *neutral* rate, or if fitness is intended only to rise, why it often was falling. Life also began with molecular selection, not cellular, but if you changed this, the theory will not add. Instead of redoing it, devices were invented. First life was explained by LUCA (it is in Wikipedia, but check). The DNA trees showed life branching at once, and evolution of molecules was a result of the *neutral* rate. If this preserved the theory, molecular selection being overlooked was ignored.

However, I do have molecular selection. Again, there was no model of it, so where did my theory come from?

Well, if an early molecule holds it sequence longer than rivals do, then the molecule spreads. Richard Dawkins mentioned this, and you can try. Long ago, before cells, if one molecule held its sequence longest for each replication, it will spread. Many early molecules spread this way. It is how evolution started. However, if molecular selection was first, but it changed to cellular, it creates other problems. The molecular selection is across RNA, DNA, ATP, ADP or ribosomes, and across life. Molecular selection does not use fitness, even if it is by natural selection. Times are different. The RNA, DNA, ATP, ADP or the ribosome will last forever. On shorter scales, the Histone H4 is across higher life. You can think that this is a 3-dimensionl (3-D) model, but it is one dimension (1-D) for the left vertical axis, but along two pathways of time. The front curve is 'real', but the back curve is along an 'imaginary' time.

Until I show this 'real' and 'imaginary' curve, nothing else that I said is controversial, of God, the Theory of Evolution, or *molecular* selection. If my facts are correct, and my theory is explained, this would explain first selection. However, my next contention, the cellular selection is along a real axis, but molecular selection is on an imaginary axis is controversial. For example, there is a *neutral* rate of gene spread, which considers both selections. However, if the gene mutation falls below the *neutral* rate, selection becomes near Darwinian only. Just if you move up the scale, to a molecular selection of about 6 x $10^{-9}$ mutations, it settles at the *neutral* rate. If you go higher, to $10^{-13}$ mutations, the gene will be 100% across eukarya, regardless of cellular selection.

The first selection was molecular. It acted on prelife, 4.1 to 3.8 billion years ago, until cellular selection began at 3.7 to 3.5 billion years. Then molecular selection was needed to explain how organisms could use cellular selection, by keeping core genes highly conserved by molecular selection. Later, there was the evolution of sex, with the addition of meiosis onto mitosis. The law of the universe that the evolution of life requires that natural selection applies along two pathways. Evolution was first used as a non-mathematical theory. When math was found, experts were reluctant to make this the entire theory. I suspect it was because no one had the equations of molecular theory.

The first part I split the facts of evolution, as is recorded, from the theory. If evolutionists agree, the problems of evolution can be solved. The next part I prove that the existing theory, without a model of molecular selection, cannot be correct. I then provide a model and include the basic math, which anyone can test. However, even with the two theories, one for *molecular* and one for *cellular* selection, there are still problems. These still form paradoxes, such as first life, the advantage of sex, or the *neutral* molecular rate.

I hope that my theories are proved to be correct, to discover how the model contained errors. You cannot have a theory of evolution that does not consider that there is molecular selection, and this applied from the start. You also cannot have a model, on a sensitive topic, which mixes up the fact and the theory. This carries into many areas, such as incorrectly drawn DNA trees, showing archaea, bacteria and eukarya evolving at once. There is also no theory of why sex is an advantage.

I hope, at minimum, these errors are overcome.

## *Facts and Theory*

This book will set out, specifically, to modify the theory of evolution. However, the book is slightly 'outside' of science. The book is scientific, but I am not practiced in science. Some things, such a full theory is harder to obtain.

Still, the first task is to split the facts of evolution from the theory. It might seem easy, but a fact and a theory are not the same, especially as a proof. A theory, say, might be a mathematical model. By contrast, facts should be data to be gathered, say, by diggings or research. If the facts and the theory align, this is fine. However, if the two do not align, something has gone wrong.

Well, facts prove that it cannot align. Right at the start, 4 billion years ago, molecular selection was first. Before early cells, molecules competed for which could survive. If there is no molecular selection as a model, that would be fine, but it seems that molecular selection was first. There are other facts as well, why it might be modeled as mathematics. To solve it, I propose a new theory of molecular selection. People can take my theory, run it, and then compare the theory to the facts. If my theory, or similar works, it can be used to solve how first selection began, 4.1 to 3.8 billion years ago, for life to start.

Except this, brings another issue.

Suppose it is true that both cellular and molecular selection evolve as a mathematical model. I will not claim it is true or not, but you cannot seem to add them! Again, this is not God's problem. If you have formulas, but you cannot add them, it is not by God. A similar problem is for the universe, except that now the theories do add. However, in evolution the same theories do not add. A huge effort is to prove that life is not from God, yet equations for the theory do not add. I wrote these equations, not God, so this objection is from me, not God.

However, the two equations do add, except to solve it, I made a guess over how evolution must work.

Now, I did not study the universe, but I studied electrical engineering. It might not seem the same, but you have all heard of the alternating and direct current, or AC/DC. Briefly, it is electrons, but the DC can add, yet the AC will not unless you change it. In evolution, if it is molecular and cellular selection, assume that the operator is genes. Genes spread from molecules evolving over huge time, and they spread if cells that compete for fitness advantage. However, if you write the equations, even as the one gene, they will not add. Therefore, no one has the equation of molecular selection. Experts must have warned that if you write the equation, the theory of evolution will fail in other effects. It is as if you must make a choice, before you write on evolution.

This then, was my guess.

I think that *cellular*, call it, Darwinian evolution, is along a real axis, which is as per standard. However, *molecular* selection, not used until now, is along an imaginary axis. Because it is a guess, it might be wrong. However, the universe does act on a real and imaginary axis. (It needs checking. I use 'j', but $j = \sqrt{-1}$. However, in the universe, $j = \sqrt{+1}$). Still, if this is mathematical, it should work. Besides, apart from other effects, there is the problem of sex.

Now, please think carefully on sex. If you are an expert in evolution, famous, you can claim it as you wish, on anything, without contradiction, except for sex. Roughly, before sex, a single organism could reproduce offspring, a 1:1 reproduction, with 100% of genes passed on. With sex, two organisms are needed to give an offspring, a 2:1 reproduction, with 50% of genes passed on. Sex is a question of why. It does not matter how clever you are, how many experts assist you, or the computers that you use, but no one can solve this.

Rather, there is no answer via a single pathway using gene selection. However, two pathways of selection can solve it. Look at Fig 0.1. If there is gain along one axis, no manipulation can explain how the front curve rises then falls again. With two pathways, though, it is easy to see how the gain in one can offset the other.

Moreover, the gene is always there with two pathways, even before sex evolved. The cutoff is the *neutral* rate, about $6 \times 10^{-9}$ mutations. For slower mutations (from $6 \times 10^{-9}$ to infinity), the gene gain goes up. For faster mutation ($6 \times 10^{-9}$ down to zero), spread falls back to Darwinian evolution. Only the two axes can prove this. The imaginary selection for molecular change can prove the 71% loss while transferring to the back curve. This theory also explains the *neutral* rate, and there is no other selectable way to explain its workings.

Apart from the gene gain for sex by math, however, there is also the genetics. Now, without the math, I am not a practicing geneticist. I can quote Wikipedia, with the math, but other effects are apparent. First, life changed from lower to higher. The book gives some details. Higher life began some 2.2 billion years ago, but sex evolved at 1.2 billion years, a billion years later. There was an increase of diversity with the advent of sex. Over those billion years, the chromosome doubled, first for mitosis, at 2.2 billion years, but then it increases again for sex, at 1.2 billion years. Remember, I can explain the math, and the two pathways of selection work, but what else did meiosis add. Well, the doubling effect allowed for diversity, but until my theory, there was no advantage for the gene to double for any other effect.

Think over the history of life. RNA evolving into DNA, the circular chromosome of lower life to the straight chromosome of eukarya, or the evolution of sex, no one thinks that doubling is an advantage. Yet there is a huge advantage, overriding other effects. If gene gains advantage on huge scales by not altering, but conserving their sequence, each doubling will be an advantage. The gains are along an imaginary axis. The math proves that if the gene halved its cellular gain by 50%, but its total gain doubled from sex, the increase is there.

Again, there is no proof, but the chromosome could not have jumped from circular to double, without an intermediate step. There must have been a set up to begin, but from 2.2 to 1.2 billion years, it doubled again. Why would it go for a billion years without diversity, but then we had a huge diversity with sex? Other explanations of sex might be spurious, if the extra chromosome was the cause.

Now, let me review the main points.

Again, the first one, crucially, is splitting of facts from the theory.

If evolutionists do this, if they split the facts from the theory, then in any dispute, evolution will win. For example, I will come to it, but there is a dispute if evolution was the cause of change on the natural world, or if the change came from God. Well, my first question is, "have you split the facts from the theory?" If they are split, it is a fact of evolution, but also a theory of how life evolved, if these are clear. However, if you cannot split it, if there is a question then evolution is at fault, for not showing both the fact and the theory.

The next task is to add molecular and cellular selection together.

The standard textbooks state that molecular and cellular evolution work together. I wish to formalize this, in a model, and add the two. If you do this, you could solve outstanding paradoxes of evolution. You can explain that first selection was molecular. You can redraw DNA trees on this. You can solve, the first time, the paradox of sex, not solved before. You can explain the *neutral* rate. You can provide a theory of molecular selection, across life. Except, if these paradoxes do exist, why does no one else notice?

Well, everyone notices.

Experts know these paradoxes exist, and have theories, but the models overlook that there is no formula for molecular selection. If anything, despite how we show it to creationists, evolution is split in two. A majority of evolutionist follow the gene-centric view. Evolution works a certain way. Someone challenges it so the experts run the equations. The equations explain many things, so there is nothing is now left to prove. By contrast, a minority concur that not everything adds. The explanation of sex, the *neutral* rates, or evolution of first life cannot be explained by the equations. They argue instead for a more holistic view of evolution. Some think this is interesting, but they challenge the holistic view to prove it via equations. Except that there are no equations, so it does not add.

By contrast, the problem is the math. While a viewpoint is interesting, if the molecular selection is missing, the total will not add. It is useless if a gene-centric view cannot solve it, but the holistic view cannot solve it either. This applies to creationism. If paradoxes cannot be solved (sex has not been solved in any time) there is no way an attitude can solve it. If the molecular selection is missing, it is better to solve this first, and let other people debate the other issues.

So, why cannot experts solve it?

There were problems in how evolution was formulated. Darwin did not use math, but a verbal model of evolution. After genes and the math were discovered, these were used, but the emphasis was to prove that the new theory backed the description by Darwin. Other controversies in the theory were ignored. The first selection was considered as LUCA. Then archaea, bacteria and eukarya, were to evolve at once. Sex remained unexplained. The *neutral* rate and modern molecular evolution remained unexplained. Experts had gone so far down an existing path that the correct answer to the problem was overlooked.

Gains of molecular selection are for domains, such as eukarya, but gains of cellular selection will be per species. There is no table of gene spread against mutation rate. Data confirms it, but experts do not allow molecular selection, so it is not in a table. If both molecular and cellular selection is by natural selection, they will follow a logistic curve. Times are from 0% to 100% and gains are 0% to 100%, but these are different physically. For instance, prelife selection was molecular, in any book. When cellular selection began, it was for archaea and bacteria, 3.7 to 3.5 billion years ago. The next selection was for eukarya, from 2.2 billion years ago. This pattern is repeated, across life.

Now consider how these add. Steven Hawking was in trouble with philosophers, for suggesting the universe is a certain way, but because he is an expert, it was fine. I have a similar suggestion, but here the experts can state that without a background in math, I cannot solve it. However, in the gene-centric view, genes give mutation, but genes provide stability, for long periods, so I wonder if these tie in an equation. At this, you might slam the book shut, because it all seems crazy. However, if the universe is similar, I wonder if evolution adds the same. I wonder if my new theory can prove how the two add.

Moreover, if you are an expert in evolution, at the highest level, for now forget about genes or biology. Instead, have a mathematician check my thesis. I give equations in Excel, but let him rerun it in MATLAB or similar. If my equations are correct, it can provide molecular selection, from the start until modern times. My theory can explain the *neutral* rate, not solved until now. It can explain sexual evolution. It explains first life by molecular selection before cellular selection. The LUCA cell is also not correct, and DNA trees must be redrawn. In biology, the success of genes is shaped as cellular selection, but an ancient force, molecular selection, is laying the pathways that the genes follow. Also, the theory works, so an overlooked effect will add.

Besides, if my theory is partially confirmed, a force like cosmology is shaping biology, and molecular selection exists from the start. There have been problems from the start, but if an issue was found, experts thought they had a theory to overcome it. Everyone in the debate tried to defeat creationists, but no one questioned if there were a problem in evolution itself. If an expert could solve this, why in many years could no one even remotely solve for sex? Why is the *neutral* rate still without a formula, decades after it was discovered?

Now, let me turn to another effect.

Over all of evolution, or any science, people can complain that the philosophers did not achieve very much. It may or may not be true, but in philosophy, they did achieve something. They worked out the difference between a fact and a theory. There were many debates. Facts are checked many ways, but for me, a fact is something you can check on Wikipedia. If you wish to know, say, why the dinosaurs died, you can check facts on Wikipedia. We know there as on asteroid strike 66 million years ago. We know it left an iridium layer at the boundary. We know some 75% of species on earth, including all the dinosaurs went extinct. This is not the theory if dinosaurs were dying from another cause. It is facts that we can check. For the philosophers it is that no matter the subject, we can never have all the facts that we need.

By contrast, the other conclusion of philosophers was that the theory of the topic was math. We can interpret it many ways, but if your equation works, that part of the theory can be checked. Today as well, the math can be checked on a computer. If the math is verified, that part of the theory is perfect. Suppose, say, an evolutionist and a creationist both had a dispute if an effect was by natural selection. Well, both people can run the math on a computer. If the results, no matter how many times it is run, give a *logistic* curve of natural selection, or a proven result, then natural selection is shown. If the math is verified as correct, you cannot dispute the result, no matter which view you support.

A strange conclusion of philosopher though, is that while formulas can be checked, there is never enough of math, to complete it. Just as, there are not enough facts to confirm each case, there is never enough math to confirm each result. It means that whether from the facts, or the theory, there cannot be an irredeemable picture of the universe. Facts and theory each have their separate weaknesses.

To check over this, please look up "evolution as fact and theory" on Wikipedia. Some quotes look wrong. To evolutionists I challenge, if you cannot do this, how do you win against the creationists?

There are many issues in evolution, from the gene-centric to a more holistic-view of evolution, from hard science to creationism, yet this issue is at the core of a science. Again, in Wikipedia, quotes contain misleading information, which I surmise as, "a fact is a theory is a fact". However, evolution is a fact and a theory, but only so that the two are compared, and then conclusions are drawn. This book will proceed on this basis. The facts are the record of how life evolved, over four billion years until modern times. The theory, where possible, is a mathematical model of how life evolved. Here, the theory can be compared against the facts. If there is a mistake, please point out the error. Maintaining that the fact and theory are one is not correct. It leads to false conclusions, and gives misinformation about evolution.

Finally, let me talk about Creationism.

Critics want to split away from Creationism, to reaffirm your faith in Darwinism. For my part, in this book, it is more important to split the facts from the theory, and only then to reach conclusions about God. If experts can split the facts from the theory, if the facts record certain events, if the theory can explain those events, evolution is fine. If by contrast, critics mix up facts and theory, treat the two as one, you will not solve the larger problems. My thesis will detail both, but then critics can compare one against the other.

To solve these problems, evolution has triumphed. To fail to solve it, but hide this under a mistruth that in evolution, a fact is a theory is a fact, is to distort the relationship of fact and theory.

# 1.0 The Basic Paradoxes

## *1.1 He Missed the Letter 'j'*

This book will give an enhanced model of evolution. Except that, most ideas in the standard theory and this are the same. Just there is a change in molecular selection, from the start of life.

The crux is that molecular selection consists not of one, but two factors of how it works. Call Darwin's theory a vertical interpretation. My view is that prior to Darwin, there was a horizontal effect, of how genes competed to spread. Assume that from the start, genes could compete by a simple means for horizontal spread. This can be proven in an equation. If a gene competes so that it lasts longer that a rival, the gene will spread, by natural selection. The two equations are:

$$\Delta x_i = x_i(w_i/\bar{w} - 1) \qquad (1)$$

$$x_k = \varepsilon_k/(1 + \varepsilon_k) \qquad (2)$$

Now, as they stand the equations work. They might look complicated, but the first is Darwinism. It produces a logistic curve. The next equation also looks different, but the curves from both equations produce the same results. This is how natural selection had first applied.

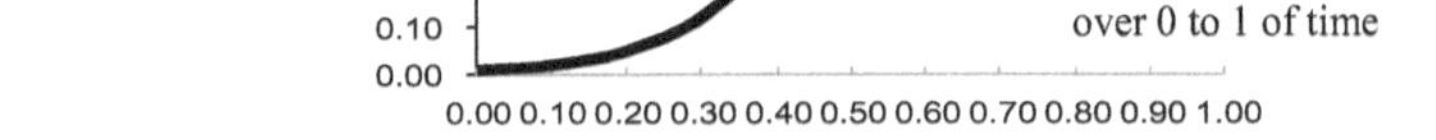

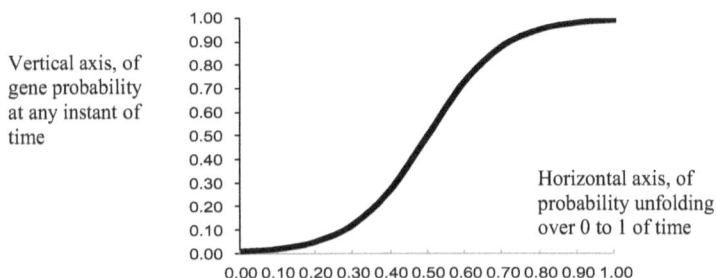


Fig 1.1.1 This is a probability, of a single vertical axis on the left. This can be unfolded, from 0 to 1 (or 0 to 100%) as time unfolds.

If there were early types of natural selection, they apply from the start. For example, most evolutionists think that the natural selection started about 4 billion years ago. Except if life began with natural selection, Darwinian selection did not apply until 3.7 to 3.5 billion years ago. There is a gap between the start of natural selection, and genes competing for Darwinism. For example, early replicators such as ADP, ATP, RNA into DNA, or the 23s Ribosome, were for life. It did not change from the first replication, so Darwinism had little effect. You might wonder, if molecular selectin was first, when did Darwinism break in?

Well, in evolution, molecular change is also prominent. There were genes, which had an effect before Darwinism split life into competing units of fitness. Perhaps the first molecules competed so they could grow best, so ADP, ATP, or RNA to DNA were used. Maybe they, and the 23s Ribosome were first. There is no difference in ADP, ATP, or for RNA to DNA. Even the first ribosome varies slightly, but these developed before Darwinism set the genes competing.

Even so, the current theory has several steps. There was the original LUCA, that developed onto branches as bacteria and archaea. Except, if you check the facts, it was not that simple. More recently there is a newer theory that bacteria and archaea evolved about the same time.

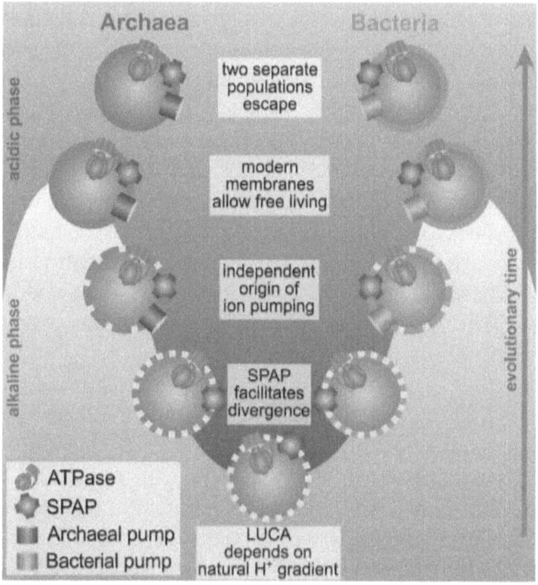


Fig 1.1.2 The separate evolution of archaea and bacteria. This is a new thesis and shows how the two formed.

If you look now in a standard book, it shows bacteria, archaea, and then eukarya in separate lines, after the first LUCA evolved. Well, there was no LUCA that formed first. Self-replicating molecules did form, early on, but to preserve early sequences. From Fig 1.1.2 (it shows LUCA, but it was not there) the split was between the archaeal and bacterial types. This was when the early types separated. Until then there is no need to split these, if genes tried to determine which could develop the pre-Darwinian types. It seems strange, but that which could hold the pre-sequence works. There is no need for Darwinism half a billion (or even to 100,000) years before the records show it was needed.

Remember, the standard theory says there was no life pre-Darwinism, but at 3.7 to 3.5 billion years ago, Darwinism began. Except, it does not align with facts. Prior to Darwinism, life was by selection, but it was non-Darwinian. If mathematicians cannot solve it, how could I without the mathematical knowledge? Well, I did study electrical engineering, which has a similar problem. I also give an equation. My equation would solve the early similarity for genes. Is it possible that there is an early effect, that the mathematicians simply overlooked?

Then, there is the letter 'j'.

Now 'j' is not yet in my equations, but it can be added. It means $\sqrt{-1}$. Often, 'i' is $\sqrt{-1}$, and 'j' is $\sqrt{+1}$, though for me 'j' conforms within biology. The issue is that this 'j' (or 'i') is at a $90^{O}$ angle to how it normally works. Strangely, it confirms how the effect operates, though it is a mathematical convention. No one says that the first axis is exactly $90^{O}$ to the second; just the math must work. For example, the axis that undergoes pre-Darwinian selection tries to keep the sequences the same. If 'j' is in the correct place, these held for pre-Darwinian evolution.

Still, suppose there is a contradiction, in that the onset of molecular selection occurs before Darwinism. No one, even among the experts, has decided a 'j' is in the wrong place. Perhaps a different explanation can do. Now, 'j' (and 'i') is used in cosmology, but why should we introduce it here? If you listen to experts, they think that the problems are solved. First life began long ago. True, there are problems, but maybe they can be solved without changing evolution.

Except, even if we can solve how life began, there is another problem, not at the start of life, but once it was three quarters evolved. What about sex? How do you explain this effect? Now, the early problems of life, say, when evolution started over which came first, can be hidden. With sex though, it was 2.2 billion years after life began, and a billion years after eukarya formed. Then, about 1.2 billion years in the past, suddenly sex formed. I challenge the reader, look at Fig 1.1.1, so how did sex work? The standard theory teaches sex must be a drop of fitness, of 2:1, because the rate of gene increase goes down by that amount. However, if you study genes, it seems to go up with the evolution of sex.

The problem is that sex cannot be explained if there is just one pathway of gene gain. However, if there are two gains, its explanation is easy. Now, my pathway is in Excel. Readers can be suspicious, because the curve is shown as a black curve, but notice there is an 'imaginary' axis that extends the total to the left. The curves can explain how sex evolved, although an explanation is needed of the result. Still, if all there is the letter 'j', how come no one else has thought of it. If Darwinism did not apply till 3.7 to 3.5 billion years, but natural selection began half a billion years before, this extra 'j' could solve the problem. Please try using this effect. It alone can then solve the problem.

Now many evolutionists, the latest is a cosmologist, argue they do not make mistakes, especially with equations. For example, the latest equation in cosmology uses a Core Theory to show how life evolved. Now, the core theory uses a component to 'j' (or 'i'), but it is correct. (There is a mix of 'j' and 'i'.) Except for evolution, the author switched to biology, but he left the 'j' out. He has left the first equation, but missed the 'j' in the second equation. Life began with natural selection, and both types can be proven. Except that evolutionists have not admitted the genes simply holding their sequence is a simpler form of natural selection.

This new change, adding a 'j' to evolution, far back at the start of life, results in a major change in evolution. Instead of one, of natural selection by Darwinism, there were two types. First, there was molecular selection, using a newer, but simpler equation. Then later, Darwinism was added. I have written both equations. Anyone, even the cosmologist, can take my equations and try them, one against the other.

If I am correct evolution has a major change. Both pathways compete for natural selection, and this continues over life.

## 1.2 The Evolution of Early Life

This book will define how life evolved as a fact. It is very dangerous. Whatever you state, especially in the early part, someone will check. Do not state effects that are not true.

This is especially dangerous, because I am presenting it differently. In the standard theory there is how genes formed, for nonlife, then in a line for bacteria, then archaea, and finally, a line for eukarya. This is a two-step process from nonlife to life, or as a nonlife as zero, then a one, two, and a three-step process.

Even at this basis, my theory is different.

First there was a zero step, as nonlife. Then there was a step to life, but as non-Darwinian selection. Here molecules sorted to survive by natural selection, but by sorting ones could grow in length. Then, 3.7 to 3.5 billion years ago, the non-Darwinian selection joined with Darwinism, for full life. From 2.2 billion to 1.2 billion years ago, life then evolved as eukarya. At 1.2 billion years ago, sex evolved.

In this, the standard theory is a two-step process, from pre-Darwinian, to full-Darwinian. By contrast, the new theory is a three-step process. The difference is by placing the letter 'j' in the second equation. If you ask, standard theory is that Darwinian selection produced the fitness, for genes to spread. In my theory, the first spread was by genes competing only to grow. Once this was laid, Darwinian selection formed. After all, if genes result from molecular and cellular evolution, it is odd if cellular evolution alone is considered. Especially, these have acted early in life. Life began about 4.1 to 3.8 billion years ago, but Darwinism did not start till 3.7 to 3.5 billion years ago. If the first selection is molecular, then Darwinian, cellular selection followed from the first.

Moreover, my math works.

Experts can complain that I am not trained. However, I have taken the split of molecular selection pre-Darwinian in Excel, and cellular selection in the first part in Excel. Both theories use natural selection, they work, but the difference is the molecular selection was before cellular selection. If I confront evolution as written, the experts have mixed the first phase up. There should be two equations, but the experts consider only one. They not only stick with this, but they do not allow evolution to continue across each phase. Again, there was both molecular and cellular evolution, and molecular was the driving force across the phases, but the evolutionists only consider cellular evolution. They did not just miss the first phase; they missed all molecular evolution.

Again, the dichotomy between molecular and cellular selection applies for evolution. The problem is that molecular selection rises with gene gain, whereas cellular selection falls as the gene gain extends. In a way, it is a guess, but in the standard theory fitness must rise for evolution to continue.

However, fitness falls for most cellular selection, so it might only rise for molecular selection. Also, it could not start with cellular selection, because genes first underwent molecular selection. Notice, if genes split between the types of selection, it began as molecular selection with a rise of fitness, but modified as cellular evolution.

Also, consider the facts. Again, experts can complain that I do not have the math to solve it, but I do have the math. Let experts, including the one from cosmology, work the math themselves. Rather, assume that the rise of cellular selection causes a fall of fitness, but molecular selection results in gains. In this case, the first acts of selection were across all molecules, so fitness rose. Once cellular selection started, fitness now fell, but this was now for separate lines. I do not prove it, but mathematically it now conforms to other facts.

Now, consider how first life arose.

There are many theories on this, almost all of them about proving that Darwinism was first, but again, it does not conform to facts. If life began with molecular selection it does not require a total enclosure, and that the first replication might have been cooler than the surroundings. After all, first life might have been hot, in which the first replication might have used energy from the surroundings to form the first replication. Remember, the first replication had to ensure that the life could replicate at all. There is little work on the first replication, but assuming it was a pool in a hotter environment. This could work.

Now, consider replication of the first types. Again, the standard theory first used trees of RNA, and used trees of bacteria, archaea, and eukarya evolving in a line. This has been modified, to show firstly, that there was no eukarya that started separately, but it arose from some combination of bacteria and archaea joining early. Remember too, there was no LUCA for this effect. My theory is that from 4.1 to 3.8 billion years it was *molecular* selection. From 3.7 to 3.5 billion years there is a *cellular* selection, but these dates vary slightly. For instance, a 23s Ribosome or ATP can evolve pre-cellular. We can allow that life had organisms to radiate heat, yet there is no reason that pre-cellular life would consist of genes and molecules to first absorb heat. It is easier to have this, so that cellular selection came later, but molecular selection was first, for prelife.

A further modification is the tree by Carl Woese in 1990, of a subunit of RNA. It shows a differentiation of archaea from bacteria of the tree, but the timing is not correct. There was no LUCA to start it, that is not correct. Then eukarya evolved about 2 billion years later than archaea and bacteria. Instead, a newer theory is that life possibly started from hydrothermal vents. Part of this is accepted, except that no one will state it explicitly. The controversy is if it were bacteria, and later, archaea, or if archaea and bacteria were together.

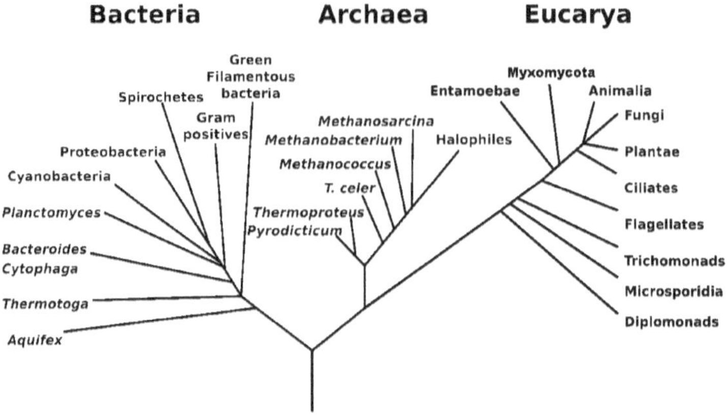


Fig 1.2.1 A copy (via Wikipedia) of the phylogenetic tree of life. Most views of life, currently available, comply with different versions of this tree.

Without the extra details, the two tree types are as shown. Fig 1.2.1 shows the phase from Carl Woese, the next, Fig 1.2.2 shows an updated presentation of how it might be. Notice that these are from Wikipedia, so we must decide which one is correct. Fig 1.2.2 shows the developments. Whatever we must discuss, we must first get the facts correct.

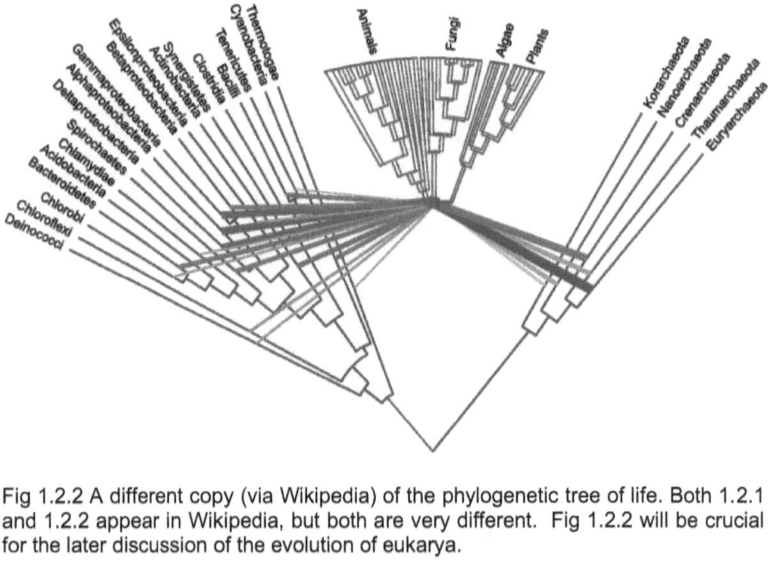


Fig 1.2.2 A different copy (via Wikipedia) of the phylogenetic tree of life. Both 1.2.1 and 1.2.2 appear in Wikipedia, but both are very different. Fig 1.2.2 will be crucial for the later discussion of the evolution of eukarya.

Next, consider the evolution of very early life. The first step is the trees by which life evolved. From here, we can divide up the process. There are various schemes for first life, but after sorting out the trees, the next step is thermodynamics. The early earth was hotter than today, and life first possibly evolved in thermal vents. Again, if we split early evolution, first life might have been simple replication, based on early selection. There is no need for an enclosing covering first, for simple replication. This could be a simple reaction, where ever (in the universe) there is water and simple replication. I do not see the need to allow more complex scenarios, if this basic requirement is met.

Now, consider the actual evolution. Looking back, we can only be sure of the 23s Ribosome, ATP, ADP and so on. There can be earlier molecules, based on an earlier type, but we have no evidence. The 23s Ribosome was formed by successive layers of ribosome molecules, the details are now in Wikipedia. We can be sure it evolved, and earlier types of this molecule can be left to research. The first molecules were by molecular selection, but even this early diagram is beyond that. Rather, even the 23s Ribosome as discussed, went through several steps to form.

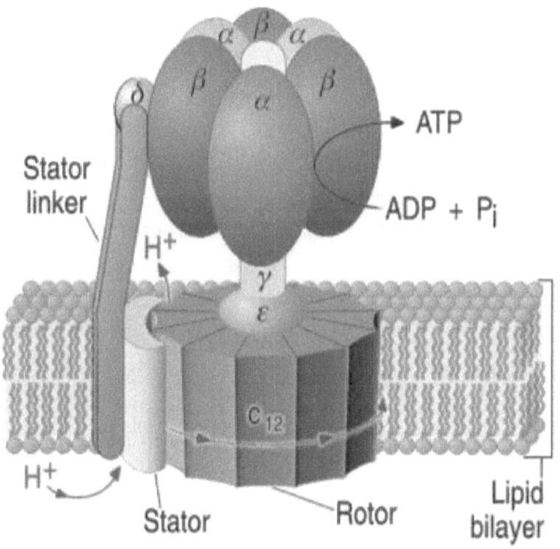


Fig 1.2.3 A simple diagram of ATP. Notice, this is already quite complex. Especially, the folding stick as ATP rotates must have evolved after the cells formed, or there is little means to explain it.

Instead, if you allow that life began with molecular selection, but it then leads to cellular selection. We can call these two axes fast and slow, but there a formal term. The cellular axis is in 'real' time, but the axis along which genes gain frequency is 'imaginary' time. There would be proteins, first 16 of them, but later it was 20. The walls of archaea and bacteria had begun. There were primitive cells, but still no selection of advanced cells. Prelife might have absorbed heat as a first stage. We cannot define this as "life" (as understood) until it could radiate heat.

Instead, the start of selection at near 3.9 (or 4.1, or 3.8) billion years ago is a fact. Evolutionists then admit that when it started, it was natural selection, but assume that it must be "Darwinian". There are such models, but none of them works. Evolutionists admit that selection must have had similar results for the first organisms, yet no one can put this together. Except, another model is possible. Richard Dawkins had one model, and it is easy to prove that it used natural selection. Rather, the issue is when pre-Darwinian prelife combines with life, as Darwinian selection. In the expert view, the types of selection cannot combine. One must cease, and be replaced by the other. The proposal is an axis of cellular selection replaces molecular selection. A prelife model and pre-Darwinian selection met with Darwinian selection.

Rather, the start of selection is a fact. Evolutionists admit that when it started, it was natural selection, but they insist that it must be "Darwinian". There are such models, but evolutionists admit that selection must have had similar results for the first organisms, but no one can put it all together. There is a caution that if they cannot begin from Darwinism, it will fail. To experts, when Darwinian selection began, my model of selection that relies on molecular selection must be replaced by Darwinism. No one can accept how both forces can proceed at one time. Before there can be life such as RNA, DNA, ATP, and proteins must have evolved. This process assumes that Darwinism must have already started.

Early prelife is also complex. Take the 23s Ribosome. If it had to convert an RNA replicate into proteins. This is a large molecule, which evolved in six stages, or more. It evolved subunits and it was circular. There must be an early, thermodynamic process, that favors molecules that form in circular formations. There is also a strong case that the first molecules were based on RNA, which was first, and is simpler in structure than DNA, which evolved later. This is from a hypothesis that life began with an RNA and Ribosome world, before transferring to a RNA, DNA, Ribosome and Protein world.

Even so, just to form the RNA world has problems. RNA is unstable by itself, is soluble in water, and so on. There are other objections, such as the early chemicals, that produce the uracil and cytosine bases prevent the adenine and guanine bases from forming. However, we must remember the power of natural selection. Evolutionists want to see "life" begun as Darwinian selection from a pre-biotic soup, but it did not happen that way. For longer than the dinosaurs evolved, evolution was by natural selection, but a pre-Darwinian form.

We know that in an early stage, molecules, which replicate accurately gained an advantage over those that did not. There were also pre-cells as a coating that formed, to assist replication.[1] John Sutherland and others have found that the uracil and cytosine bases could form, but also be assisted by intermediates for the adenine and guanine bases. Over 500 – 100 million years before life, there can be many intermediates. Prior to proteins, structures were derivations of ribosomes. It does not require pure RNA, and there might have been early DNA, or other ribosomes, and molecules simply trying to copy accurately. There was no fitness of cells competing, which did not exist, and early molecules might have absorbed heat, rather than radiated it.

If anything, if we take the first replication back to the start of prelife, there are many possibilities. Again, there are difficulties modifying pure RNA, so maybe it began as mixtures of RNA, DNA, or other molecules. There have also been RNA base structures seen in interstellar objects. Perhaps part of the early formations was to use these, in early Earth. Remember, the first goal is to grow molecules longer, and it evolved over hundreds of millions of years. Perhaps at 3.9 billion years molecules began to replicate, but then millions of years later, RNA and DNA were achieved. That replication was for conservation, and RNA can retain conservation for billions of years, supports this view.

Again, "my theory" is that the evolution of sex is a result of selection having to modify along real and imaginary axis, and I will stick to this. However, Nick Lane and others have suggested a different problem, that the conflict was the evolution of mitochondria, essential to the evolution of eukarya. To illustrate this, they have shown the evolution of early life, more detailed than I (though I include it). Until now, I held the standard version of early life that it occurred as per Fig 1.2.1. However, the split of archaea and bacteria is now detailed, and is to be used instead. The split for the ATP module is also given. In less noticeable parts of Wikipedia, there is a discussion of how the archaea and bacteria evolved, with an emphasis on how the gradients were set up. This is oddly a different theory than my emphasis on the gene setup, but worth pursuing. Nick Lane and others have suggested that archaea and bacteria originated around the same time. This shows a tree differently as in Fig 1.2.2.

In my model, the theory follows a logistic curve, and moves from 0% to 100%, if it is successful. Except, cellular selection also moves from 0% to 100%, but the curves are not the same. No one will admit this, but it is where the standard theory strikes a problem. Anyone could formulate a theory for molecular selection, but if you add it to *cellular* selection, this becomes disjointed. This is where results do not add. My solution is that

---

[1] Jack Szostak and others made the point, that these linkages favor increased replication, for the generation of pre-cells. Jack Szostak showed how various combinations were tried. Remember, if the goal were longer RNA and DNA, conservation for sequence would help.

in the theory, the first, molecular selection adds along an imaginary axis. (I suggest that the universe works a similar way, but critics can object that I cannot do this, because it is so unfamiliar.)

Again, the basis of my theory is a split, early in life, between molecules first selecting for non-Darwinism, then later, for full life, Darwinism and pre-selection combined. Experts can object that I cannot do this, I must be wrong, because the doctrines state clearly that first life evolved as one unit. My objection is that the split is clear, supported by facts. Life began about 4.1 to 3.8 billion years ago, and was selection to form the first molecules. Then it was joined by full Darwinism, between 3.7 to 3.5 billion years. These are separate equations. The new equation does not use Darwinian fitness, but it will progress by natural selection.

So, how does it work in an equation?

Simply, you must add the letter 'j' to the second equation. If you add 'j' in Equation 2, it suddenly works. The molecular selection will stay the same, right from the start. In this way, even if Darwinism did not enact for the first half-a-billion years, early replication would not be affected. Prior to Darwinism, the act of selection could sort the first molecules, to half-a-billion years, before Darwin's selection took hold.

## 1.3 Sexual Evolution

The most famous paradox in evolution is that of sex.

If a gene from lower life, archaea or bacteria, reproduces, it duplicates its genome, a 1 to 1 reproduction, with 100% of genes passed on. If a gene in higher life, eukarya tries to reproduce, for the first billion years it could, via mitosis. After a billion years, sex was added by mitosis adding on meiosis, so now two parents were needed for an offspring. It was a 2 to 1 reproduction, with 50% of genes passed. If eukarya can first reproduce without adding meiosis onto mitosis, why is this extra change needed? There are theories, explanations, books, and exceptions, but the paradox of sex is a question, why?

The book has a new explanation, not considered before. First, there is a gain along the far grey line, with addition of *molecular* selection onto *cellular* selection. Next, it was not noticed, but there is a major change when sexual selection adds to the diversity of the clades. Neither of these effects has been easy to work out.

Now, if archaea or bacteria could reproduce by a 1 to 1 reproduction, why did the higher cell start? It seems that archaea needed to reproduce by enclosing bacteria. Instead of two reproductions, archaea had now to include bacteria, so both types could reproduce as one. The new type, eukarya, not only reproduced as one, but it lived, gave birth, and died. Once the two combined, the separate circular genomes divided to appear as linear-type chromosomes. It started 2.2 billion years ago, but persisted until 1.2 billion years ago, when sex was added.

In that case, what advantage did sex bring, not there before?

Well, there is a huge increase in diversity at 1.2 billion years, when sex evolved. Then, some six new clades and several minor ones arose. The groupings also seemed to branch quickly. However, at that time there was a drop of cellular gene spread, from 1:1 down to 2:1. How could a sudden drop in cellular selection, to 50%, at the same moment cause an increase in diversity.

Instead, there are two types of selection. If you combine molecular and cellular selection, a combination does produce a gain. You are free to take these two types and test it, and notice how with the two gains, an increase for sex is possible. On the new curve, regardless how the front curve responds, the back curve will increase if the mutation rate stays high. If a gene for eukarya starts with a $10^{-13}$ or even slower mutation, it will spread across higher life. If, once sex has evolved, a gene increases mutation up to $6 \times 10^{-9}$, the gene will also spread in higher life. If the gene can increase its mutation faster than $6 \times 10^{-9}$, it will quickly lose direct gain. At this point for mutations more than $6 \times 10^{-9}$, the new gene will follow standard gain at the front. Roughly, the curve in Fig 1.2.2 existed from first eukarya (possibly *diskagma*).

This would have existed after sex evolved, for many types. It seems that the first eukarya was able to reproduce, but it only allowed production in a narrow line. Remember, the back curve allows a gain for sex to spread across many types, but the first one did not widen. It seems as if once the first eukarya evolved (if it were *diskagma* or similar), it could reproduce the early eukarya, but this did not widen. A rate of mutation above $10^{-13}$ (about) lets the gene spread widely, but only for the first eukarya. However, then sex came, the gene could spread all the way down to 0.5 to 0.71 % of the fitness (see Table 2.6.1). Suppose that archaea and bacteria joined some 2.2 billion years ago. The first eukarya spread, but nothing else branched until 1.2 billion years, when sex was added. It seemed it was sex, rather than anything, broadened the clades.

Again, please check Fig 1.2.2 and so on. There is no way the curve in Fig 1.2.1 can produce the gain for sex. Even if not an expert at math, there is no manipulation of the single curve to do it. There must be two gains, in which the back curve is the total mutation rate. I have been questioned about this, because by using $x_i$ alone the gene seems more conserved with a higher mutation rate. However, it is the lower mutation of the far grey line that gives the gain for sex. The other explanations, no matter how researched, cannot produce a result of sex.

The other problem is that even after sex, genes engage in a horizontal transfer, which has no explanation in the standard theory. Manipulations of the front curve alone, no matter how clever, there is no mechanism of why the gene goes from the front to the back, by any type of horizontal transfer. However, with my theory it will. Check Fig 1.2.2 again. If there is a gene which evolved after sex, but it needs to migrate to the back curve, it can still 'horizontally transfer' by leaving the normal after sex route, and directly transfer to the back curve (it can appear as a dotted line) by moving outside of normal sex. Notice too, all the other theories of sex use math, but assume that the gene spreads along a Darwinian, cellular selection. None allows that the gene must spread along two axes, molecular and cellular. The previous theories try to prove how a fall of cellular selection from 100% to 50% was a gain. Instead, cellular selection was from 0% to 100% for any species, with or without sex. Molecular gain was 100% for the first histones, but it fell from there, to 71% for full gene spread, as a *neutral* rate. If gene selection falls below 71%, moreover, it will revert mostly to Darwinian spread only.

To evolve from archaea and bacteria into a eukaryote was complex. The genome had to increase about 1,000 times, and change from a circular chromosome into the double chromosome in eukarya. There was also change, so that copying in lower life ceased, to become a parent-child set up. However, the big change was by mitosis mutating to include meiosis. Mitosis might have started at 2.2 billion years, but meiosis was added at 1.2 billion years when sex was observed.

Again, I have no proof, but suppose the chromosomes doubled with sex, but the autosome (of one gene) chromosome stayed separate. This continued until the allosome (for sex) divided into species, which would lead to male and female. This first occurred when archaeon ingested a bacterium. The two duplicated before the cell divided, and was about 2.2 billion years ago. When meiosis was added, another split occurred at 1.2 billion years ago. The chromosome could not jump from circular to a double chromosome, without some intermediate steps. Once the first ingestion occurred, separate chromosomes were needed, perhaps one for archaea and one for bacteria, but with a way to replicate. Splitting of the chromosome in two allowed this. A billion years later, extra complication needed the double chromosome, by allowing sex.

Now, in the standard theory there is a dispute if meiosis added onto mitosis after a long period. Again, no one seems sure, but an assumption is that it took a billion years for the mitosis and meiosis grouping. We cannot be certain when the second chromosome joined, but only that sex evolved about 1.2 billion years ago. It is hard to see how mitosis could be a long time, but meiosis results from a different cause, such as bacterial division. More likely, bacterial genes diversify by horizontal exchange, but this was not possible once pre-eukarya evolved.

From a first four-fold reproduction, eukaryotes could start to branch, or those that emulated a four-fold reproduction survived. Other attributes of sex, where isogamy evolved into ansiogamy, DNA repair, a two-fold cost, or loss of sex in a few species, were later. There are other theories of sex, including the two-fold cost, resistance to parasites, genetic variation, deleterious genes, novel genotypes, DNA repair, and speed of evolution. Suppose that experts, even if they have written about sex, even if they have tried the other equations, still confess that really, they do not know why sex evolved. Again, the answer is to try my model.

## *1.4 Human Logic ...*

Why can the human brain use logic?

This is a key question, for which I have a theory. Again, this theory is not mathematical, but we must consider this. There is a point where the self-reflecting neurons inside the brain, cut themselves from the outside. This is where the 90 billion neurons and trillions of neural connections become self-reflecting. Someone can calculate this. If this part of the brain is self-reflecting, this is defined by logic.

Now, let me quote from Darwin. If you write on evolution, no matter how correct you think that you are, do not contradict Darwin. If Darwin is correct, there are other means to make the point. However, on the brain, which is crucial to the arguments, Darwin wrote

> The difference in mind between man and the higher animals, great as it is, is one of degree and not of kind. **Darwin 1871.**

Now it is said, I put the question to evolutionists about the mind of man. Is it the human brain the same as other animals, varying by degrees as needed, or is it different?

Perhaps all evolutionists have the same answer, that once variations are taken, that the mind of man is essentially the same as other animals. If it formed from the same biology, it is subject to the rules of evolution. If it is flesh and blood, how can it be different?

Well, it is not just different, but a different device in the universe.

This has nothing to do with biology, genes, fitness, or any of these. Instead, the difference is logic. An extra growth in the human brain left it free to do logic. Yes, it is a function of biology, but the human brain, or some brains, became internalized. You need facts, animals use them, and facts can be connected. Yet for logic, you do not need facts. If a brain can grasp that $1 + 1 = 2$ independently of the objects that these represent, with assistance, it can work that $1 + e^{j\pi} = 0$ by a similar rule-driven process. It is like Einstein working out that gravitational waves exist, a century before they were proven. It is like Newton, figuring that the force that pulls apples to the ground must also hold the planets in their orbits. This is independent of how evolution unfolds.[2]

If you can grasp that the human brain uses logic, and it is independent of other factors in its evolution, you have grasped the human condition. It is useless to argue that because of some attributes in our evolution, we must think a certain way, when the finest human thoughts are of logic. It

---

[2] There is an entirely new theory here. The issue is that there is not only zero genetic cost to abstraction. When scientist began designing quantum computers, they realized that in theory there is zero energy cost to analytical thinking, in which information is conserved. (Note, the principle was noticed via quantum computers, but it is not because of them. This is another example of basic principles that everybody should have thought about long ago.)

is useless to argue how we are trapped into behaviors, when the point is to raise our thinking above such levels.

Interestingly though, the human brain is logical, from evolution, so how did it get there? Hominids could have done all that they needed to do, hunt, survive, procreate, migrate and multiply with a smaller brain, and many did. Everything evolves at a cost. If a fish lives in darkness, it atrophies use of its eyes rather than pay the fitness penalty to retain those organs. Yet the brain, with 2.5% of human body mass, consumes 20% of the body's energy. Rival hominids with smaller brains had more energy for hunting and producing offspring, and those offspring were easier to bear because the head was smaller. Yet the rivals perished. To survive, be fit, and pass on DNA on the plains of Africa, reproductive fitness was not just measured in total offspring or catching prey. The brain needed to be capable of logic, as mathematics, music, and philosophy. There seems to be no simple explanation of why.

My hypothesis is that neural circuits used for leaning offer the most advantages for rapid brain expansion.

When brains first evolved, each neural circuit had to be designed by selection, a tedious process that perhaps required a new gene for each new circuit. There are advantages in 'fixed' circuits, because they are fast and reliable (like fixed logic in computers), and all brains use core logic for crucial functions. The drawback with fixed neural logic though, is that it will quickly consume the available number genes to encode it, if each circuit needs one gene. The evolutionary solution was 'learning' circuits. These can be of a common genetic design, but they can be wired after birth by experience.[3]

There is also a larger frontal cortex in humans, which occupies 29% of the human brain, but 17% of a chimp's brain. The cortex is a conversion of functions that were earlier performed by reflex. However, the frontal cortex is a further conversion of functions of the middle cortex. This gives a ratio of frontal cortex to reflex of 50% in chimps, but 230% in humans. This way the human brain could possess almost a 99% commonality of circuit design with a chimpanzee, yet it can still be a radically different brain because of the different ratios of reflexive neurology, to higher and prefrontal cortex neurology.

When humans evolved, they needed to learn a huge range of new skills quickly. Walking was still in the hindbrain, but now supplemented by a vast expansion of learning circuits. (So while crawling is mostly reflexive, for humans walking upright is a learned skill.) Many other human skills, such as speech or tactile dexterity, are also in the learning cortex. Even vision is partly learned, and it makes sense. To perfect the new skills that

---

[3] This is oversimplified. All neurons have a slight amount of learning by synaptic facilitation, and even learning circuits can be of 'wire-once' or 'wire-many-times' types.

humans had to learn by selecting a new gene for each skill would have taken many genes and a huge evolutionary effort. However, once the design of a learning circuit is perfected, these simply need be multiplied millions of times for the effect. [4]

The human brain can have three times as many neurons as a chimp, or seven times more than a mouse, for a like number of genes in each species. See that the human brain case expanded in volume by a genetic instruction. However, the increased volume was not filled with individually designed and selected neural circuits.[5] It was filled with learning neurology, for an increased brain size and capability, but a small genetic cost of change. This indicates relative expansion of the generalized cortex against function-specific parts of the human brain, compared to a chimp. The expansive non-shaded area does not mean that the human brain is *tabula rasa*; a so-called 'white paper' as philosophers once thought. If humans do not study mathematics or play chess, it does not mean that their brain is empty. Research has shown that the brain is used, with most of it for naturalistic functions. These include activities such as walking, speech, or recognizing faces.[6]

Yet why did the human brain still evolve so large, to an extent that it could grasp mathematics or logic?

Brains had been growing larger for half a billion years, when human evolution began. The reason was that brains made behavior flexible, so that organisms could adapt at little cost of total genetic change. Learning neurology furthered this trend, by allowing greater adaptation for fewer alterations to core sequences. Once mammals evolved primate brains grew larger, and hominid evolution triggered a competition among individuals for increased brain size was a way to adapt.

I believe that out on the plains of Africa when the human frontal cortex expanded that extra few cc, something unforeseen happened in the history of life, and perhaps in the universe. Everything evolves at a cost, and the human brain has evolved not one more cc in volume or one more neuron than it needs for the survival and reproductive needs of the individual.[7] Yet statistically, brains still vary in size and neuron count, and different learning experiences can dramatically increase the number and complexity of neural connections.

For 90 billion neurons and trillions of neural connections, perhaps the human brain sits at a critical mass of neuron count to learning ratio. Below

---

[4] Tragic evidence for this was discovered when a child lost the sight of one eye when that was bandaged for a minor ailment for the crucial few weeks when the eye "learns" to see.

[5] This diagram shows differences between functionally specific and generalized areas of the higher cortex. (Seeing as the brain stem is reflexive, I have colored this gray at my own discretion, but it was white in the original diagram.)

[6] We have learned from computers that it is harder to recognize faces than play chess!

[7] I use "plains of Africa", but it is possible that the final tribe evolved closer to the shoreline or at least along riverbanks (though nothing to do with the "aquatic ape").

that mass, the brain is back in the plains of Africa, struggling to survive, hunt, form alliances, and procure offspring; for selective reasons. Above that mass the brain leaves Africa, evolution, offspring, and hunting. Like Newton, "Voyaging through strange seas of Thought alone",[8] above a critical neural mass the human brain both cuts itself off from the universe, but also finally becomes a part of it.

The 90 billion neurons and the trillions of neural connections, with an appropriate learning process, become a critical mass of firing states of synapses that *internalize* the thinking state of from the external world that surrounds it.[9] From internalization of the firing states comes abstraction, and the higher mental activities, including speech, logic, mathematics, and philosophy. Once enabled, abstract thinking consumes metabolic costs, but zero genetic cost to evolve further. This final critical balance of the human neural mass might have a great advantage.

For four billion years, life has selected to adapt complex organisms into new modes of existence for smaller changes to core genes. Humans evolved from an ape-like ancestor for a cost of 1.5% change of genes, and fusion of two chromosomes. That is dramatic efficiency. Moreover, the human species can adapt into almost any niche on Earth, or even explore outer space, for near zero change to the genes that the species migrated with from Africa.[10] That is also a dramatic efficiency. The most dramatic change though, is that the human brain can alter its state from a reflexive organ of survival to one of abstraction for zero alteration to genes. That is perhaps the maximum efficiency that the evolutionary process might achieve.

Nevertheless, before *Homo sapiens* emerged, an extraordinary event occurred in evolution. New biological organs, such as vertebrae, feathers, or a four-chamber heart, use large changes in genomes to evolve. This in turn requires time, mutation, and huge ecological and selective pressures to cause the changes, which is why most changes occurred in the deep past, when such times and pressures were available. Here, 50,000 years ago, a miniscule change in DNA resulted in the evolution of not just a radically new organ, but also a type of device in the universe. The hominid brain became a human one.

---

[8] This is quoted from Wordsworth. In earlier books, I referred to how Steven Hawking could ponder the universe from within a crippled body, but was told that this might be offensive. Still, I think that Steven's own quote from Shakespeare; "I could be bounded in a nutshell, and count myself a king of infinite space", states the principle anyway.

[9] The first cell was a form of information barrier, inside which internal order can increase against thermodynamic probability. The synaptic mass of the human brain might be a further information barrier, which allows large increases in order for little genetic change.

[10] Again, the 'genetic distance' between the original African races and that of the migrated species is very controversial and hard to measure for original populations anyway, because of subsequent mixing. Migrated species had to adapt to different, often more seasonal and more challenging environments, which produced some difference.

In summary, it was not that a few alleles altered, and as a result, humans evolved a large brain, to employ speech. Instead, brains have been growing larger, and higher animals have increased intelligence, for many reasons over the history of life. One reason is that when life was simple, genomes could afford to alter a large amount to adapt, because life can replicate in huge numbers, so fitness rewards were high. However, as life evolved in complexity, there was intense struggle to reduce the amount of alteration needed to adapt to new conditions. Versatile behavior in animals allows rapid adaptation for low genetic cost. Large brains with a high ratio of learning to reflex have furthered the trend. In hominids this continued until an organism evolved that could adjust its thinking by internalizing the thought process itself, without further genetic change.

The human brain grew for many reasons. Yet it did not grow towards a goal of logic or abstraction, because these traits are rule-driven processes, which would be hard to select. Instead, evolution unfolded by its own processes, and the brain grew to enhance fitness of its possessors. During hominid evolution a point where the balance of learning to neural density led to a critical mass of internalized firing states in the human brain, from which abstraction and intelligence evolved beyond that.

# 2.0 Modifying Evolution

## 2.1 Life Evolving as a Fact

How did life evolve, as a fact?

We do not yet state why it evolved, perhaps it was by God, but we do want to be sure of the facts, we must explain.

Before we start, facts are well established. It is similar to the recently discovered gravitational waves in physics. Einstein found the equations for these 100 years ago, but doubted if they would be found. Recent tests, though, in the USA, and elsewhere, confirm their existence. It is useless to argue that the waves were not there, on ethical or religious grounds. Unless you are an experimentalist in that area, or have extra knowledge, accept the fact as it is presented.

Similarly, the most important facts of evolution, not the theory, but the gross facts, are checked across many disciplines, such as astronomy or geology. One key fact, against modern life, is how old the Earth is. Yet other facts, within an overall perspective, is how small the time on Earth, against the age of the universe. All these facts are checked again, many times, across many disciplines.

First, the universe is near 13.8 billion years old. The sun and planets formed some 4.6 billion years ago. There was a late heavy bombardment over 4 billion years ago, and life on Earth can have started shortly after, 4.1 to 3.8 billion years ago. It does not change the facts, but for clarity, my theory is that non-Darwinian molecular selection would have begun for prelife at 4.1 to 3.8 billion years. Darwinian selection began at 3.7 to 3.5 billion years ago.

Again, if you were thinking that Earth was formed 4,400 year ago by God, 13.8 billion years for the universe, or 4.6 billion years for the Earth seems a long time. However, the Earth might still be young. Rocky type planets, such as Earth, must come from Population I suns. This means Population II suns must form first, and then disperse to form Population I systems (check details). If life on Earth lasted another billion years, it is still a short window on the age of the universe. If life in the universe can last perhaps a trillion years, a billion years in future on Earth is short. We imagine advanced civilization out there, perhaps with communications between stars. Well, it may be possible, more as civilizations evolve. Just, we have no evidence of other civilizations yet, so perhaps this lies in the future, beyond life on Earth.

In any case, let us return to life on Earth. We know that it exists, so how can we be certain of the facts of its evolution?

Well, again, life on Earth took time to evolve. Darwin spoke of early life, but this was at most half-a-billion years. Instead, life, with prelife, evolved for about 4 billion years. From a distance, another planet say, most early selection would not be observable. Multi-cellular organisms did not evolve until about 1 billion years ago, but animals did not evolve until 620 million years ago. At 530 million years, the first footprints on land were recognizable, and at 434 million, primitive plants moved onto land. By 363 million years, living earth is recognizable from space. However, this was less than 400 million years. Rather, what happened over the huge time when life was developing?

Again, there are new disciplines, such as paleobiology that deal with this. Fig 2.1.1 shows the overall effects.

The graph shows life starting at 4.1 to 3.8 billion years ago. The pre-Darwinian phase went for 500 to 100 million years. Life started at 3.7 to 3.5 billion years. Photosynthesis from cyanobacteria was at 3.0 billion years, and an increase in oxygen is from 2.4 billion years. From 2 billion to 1 billion years ago, we have a "boring billion", as it has been called. There were glaciations, 2.4 to 2.1 billion years ago. The evolution of higher life started at 1.8 (I think it was 2.2) billion years ago. Evolution of sexual reproduction was at 1.2 billion years ago. It was followed by evolution of multi-celled life at 1 billion years, until there were animals at 620 million years. This period included another snowball earth, at 850 to 630 million years. Snowball earths might have covered the planet, but possibly parts near the equator, or at hot springs, allowed for water.

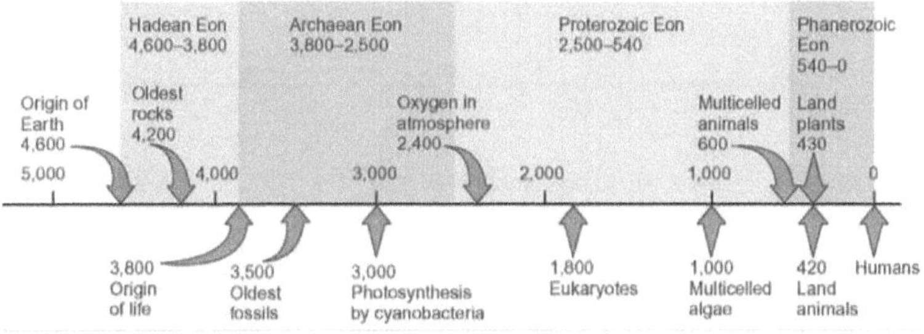


Fig 2.1.1 A tracing (modified from Wikipedia) of the rates of evolution. This shows the huge times involved, especially against the late appearance of land animals, and the very late appearance (almost 0 on this) of humans.

Another problem is the rise of oxygen levels.

The evolution of higher life might have taken two billion years, yet this was also tied to the oxygen levels. For instance, did evolution of the double walled bacteria occur once there were higher organisms to receive these? Perhaps 2.2 billion years ago double celled bacteria evolved, but it is not clear. This came with the evolution of higher organisms, where ancient life modified to fill them. Prior to oxygen, *thermophilic* organism metabolized sulfur, directly converting heat to energy. This allowed for different types, although this limited life to heat and sulfur sources. Cyanobacteria could move away from these heat sources by converting sunlight to energy, releasing oxygen.

As oxygen levels increased, it allowed new organisms to use oxygen as a rich energy source. Still, there is a cost to evolving new types, so organisms try to avoid paying it, if they can adapt by simple changes. Another explanation of the stepped effect is that the geological record exists in layers; of sameness punctuated at a boundary. A stepped change in geology (as rocks) can partially explain its mirror in sudden change in the species (fossils). In this, life itself has interactively helped mold the geological record, as now shown. There are different ideas of geological changes occurring as periodic catastrophes, such as asteroid strikes, or even a periodic 'death' star, or heightened volcanism for the steps.

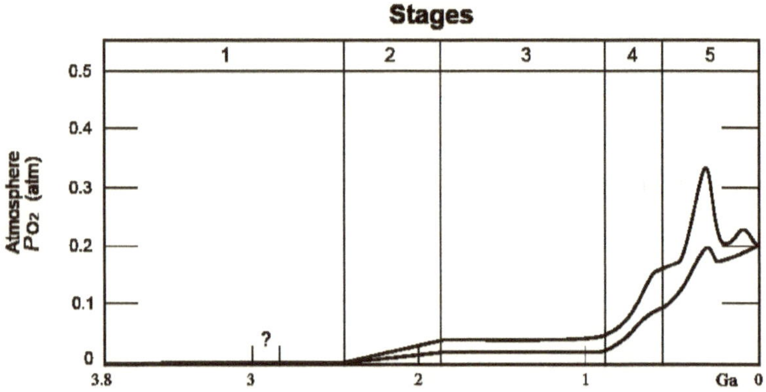


Fig 2.1.2 Tracing (from Wikipedia) of maximum and minimum oxygen levels over the history of life. Maximum was for a period of large insects and amphibians, at the end of the Carboniferous period.

Finally, we must consider the "filling up" of available space. If we look at modern life, everything is filled, but why did it take so long? Before 600 million years, even the oceans did not contain large organisms, and until 430 Ma (million years ago), the land was not filled. Even lower life, could not fully fill the space. It is only in modern times, when the sea, land, and air are filled, do we reach saturation.

This is where the punctuated rates begin. At 580 million years, the Ediacaran began. This part was followed by five major extinctions. At 450–440 Ma, was the Ordovician-Silurian transition. These two events occurred that killed off 27% of all families, 57% of all genera and 60% to 70% of all species. Together they are ranked as the second largest of the major extinctions in Earth's history. At 375–360, Ma was the Devonian-Carboniferous transition. In the later Devonian, a prolonged series of extinctions eliminated about 19% of families, 50% of all genera and 70% of all species. This extinction lasted for 20 million years, and a series of extinction pulses within this.

At 252 Ma, there was the Permian-Triassic transition. Earth's largest extinction killed 57% of all families, 83% of all genera and 90% to 96% of all species (53% of marine families, 84% of marine genera, about 96% of marine species and an about 70% of land species, including insects). The successful marine arthropod, the trilobite became extinct. The evidence of plants is less clear, but new taxa became dominant after the extinction. On land, it ended the primacy of mammal-like reptiles. The recovery of vertebrates took 30 million years, but the vacant niches gave an opportunity for archosaurs to become ascendant. The late Permian was a difficult time for marine life, even before the "Great Dying".

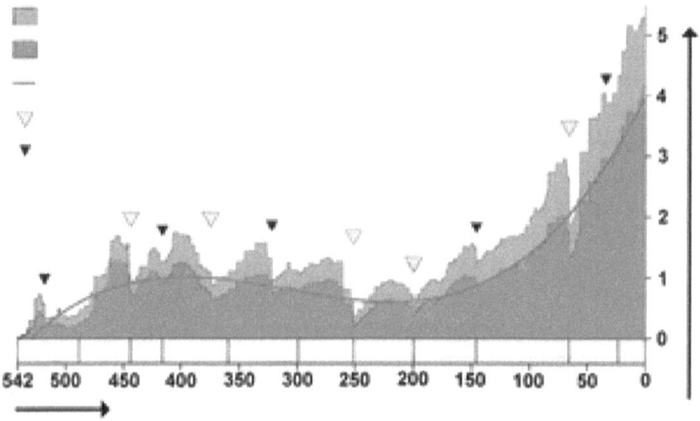


Fig 2.1.3 Copy (from Wikipedia) of the major extinctions. (Heavy line shows major decrease, light line shows general decrease.) Humans too, even on this scale, would be barely visible in the last part, about 300 to 150,000 years.

At 201.3 Ma was the Triassic-Jurassic transition. About 23% of families, 48% of all genera (20% of marine families and 55% of marine genera) and up to 75% of species went extinct. Most non-dinosaurian archosaurs, most therapsids, and most large amphibians were eliminated, leaving dinosaurs with little terrestrial competition. Non-dinosaurian archosaurs then continued to dominate aquatic environments. Non-Archosauria diapsids continued in the marine environments. The Temnospondyl lineage of large amphibians survived until the end of the Cretaceous in Australia, before vanishing.

At 66 Ma the Cretaceous-Paleogene (it was the Cretaceous-Tertiary) occurred. This has been related to an asteroid strike, at 66 Ma, which may have caused the decline. About 17% of families, 50% of all genera and 75% of species became extinct. In the seas it reduced the percentage of sessile animals (those unable to move about) to about 33%. All non-avian dinosaurs became extinct during that time. The boundary event was severe with variability in the extinction between and among different clades. Mammals and birds, the latter from theropod dinosaurs, emerged as dominant large land animals.

Next, consider the actual level of change.

Now, one of the main facts for evolution is that no fossil has ever been discovered out-of-sequence. Recently, there has been a dramatic proof of how it might work, from the Evo-Devo development by Sean B Carroll, and others. An example is hox genes. It was thought that if the eye developed in different ways, perhaps there were developments for different eyes, say, for fish, insects, and mammals. Evo-Devo proved that eyes are controlled by a position of hox genes, common across species. Eye development is also controlled by a PAX6 protein, common such the human and mouse, or even mouse and fruit fly genes can be exchanged. The common genes are from way back, at least 620 Ma ago, and the hox at least before the fungi and animal split.

Another example is that humans and fruit flies appear very different. Yet again, both have hox genes, of almost identical sequence at the micro level. At a higher level, though, fruit flies have a disjointed cluster of ten genes spread over nine places. Humans have four clusters over thirteen positions (see Fig. 2.1.4 next). Creationists have contended that there is no link between an invertebrate and a vertebrate. However, a modern amphioxus and a vertebrate share a near identical set, to the first ten hox genes. In early vertebrates, the hox set has duplicated, and for modern fish and land vertebrates, it duplicated again. When a vertebrate diverged from an invertebrate, its micro level DNA sequences diverged roughly linear in time. However, not micro-level, but macro changes in hox clusters result in the difference between an invertebrate (1 cluster), an ancient (2 cluster) and modern (4 cluster) vertebrate.

If anything, gene duplication, rather than changes of alleles, seems to be a source of step change.

For example, genes of the globulin super-family in vertebrates have undergone many duplications in 500 Ma, but with micro-sequences less diverged than macro-ones. The divergence between β-globulin genes in humans and mice is less than a ε-globulin divergence between β-globulin and ε-globulin genes, in either humans or mice. The macro-event was the β-globulin and ε-globulin split, before humans and mice diverged. Since, the evolution of the mice and human lineages was a micro-event within established globulin families. The chromosome, duplications of a globulin gene distinguish reptiles from mammals, but micro-changes in a globulin gene distinguish a mouse from a human.

However, while certain macro-mutations such as gene duplication can be viable, there is no theory of how they are maintained across species. Yet gene duplication allows the original not to alter its sequence to spread in other populations. Instead, via duplication, the original genes are able to obtain the variation that the genome needs to adapt in many lines, by the duplicate altering to adapt. For instance, for the micro-level organization of genes, humans are closer to chimps, than chimps to other great apes. It seems odd, but the macro-level, humans have only 23 chromosome pairs, but the other great apes have 24 pairs. Chromosomes 2 and 3 fused in human evolution, and others inverted. Genes of the four species of great ape are similar. One level up, however, the chromosome makes three of those species an ape, and just one of them a hominid.

| *Position* | 1 | 2 | 3 | 4 | 5 | 6 | 7 | 8 | 9 | 10 | 11 | 12 | 13 |
|---|---|---|---|---|---|---|---|---|---|---|---|---|---|
| | | | | | | | | | | | | | |
| Fruit Fly | x | x | x-x | x | x | x | | | | | | | |
| | | | | | | | x | x | x | | | | |
| Amphioxous | x | x | x | x | x | x | x | x | x | x | | | |
| | | | | | | | | | | | | | |
| First Vertebrate | x | x | x | x | x | x | x | x | x | x | x | x | x |
| | | | | | | | | | | | | | |
| Human HOXA | x | x | x | x | x | x | x | | x | x | x | | x |
| Human HOXB | x | x | x | x | x | x | x | x | x | | | | |
| Human HOXC | | | | x | x | x | | x | x | x | x | x | x |
| Human HOXD | x | | x | x | | | | x | x | x | x | x | x |

Fig 2.1.4 Simplified *phylogenic* evolution of hox gene clusters. Each hox gene (shown as an "x") is 183 bp long and has a mostly similar sequence by position. Yet the macro organization of hox gene clusters is very different. The hox gene arrangement of the first vertebrate is only hypothesized, but is likely.

Gene duplication, changes to chromosomes, or any large genomic events, can be called macro-mutations. Still, there are reasons that this term is not favored, to do with scale and genomes reflect this scale.[11] Changes at the micro level should be for one bp to give selection time to act. Large changes, like inverting or duplicating a gene, or chromosome, should occur as complete units. However, a rash of mutations at the micro-level, without structure, does not seem viable. [12]

Now, consider human evolution. This might not seem central, but it is a focus. Studies show that primates diverged from other mammals about 85 Ma, in the Late Cretaceous period. The earliest fossils appear in the

[11] In a car, if you made a macro change, such as moving the engine to the back, it might improve the car. If you attempted a micro-change, though, such as moving the crankshaft to the back and leaving the engine block in the front, it would not work.

[12] Base pairs (bp) are read in threes, in a codon. Altering one or two bp (a point mutation) can be a neutral change if it does not affect the codon structure. Dropping one or two bp can cause catastrophic disruption to the reading sequence (a frame-shift mutation). Dropping or inverting a codon is a macro-mutation, which might not disrupt the reading sequence.

Paleocene, around 55 Ma. In the Hominoidea (apes) superfamily, the Hominidae family diverged from the Hylobatidae (gibbon) family some 15–20 Ma. African great apes diverged from orangutans (Ponginae) about 14 Ma. The Hominini tribe (humans, *Australopithecines* and other extinct biped genera, and chimpanzees) parted from the Gorillini tribe (gorillas) about 8 Ma. The subtribes Hominina (humans and biped ancestors) and Panina (chimps) separated about 7.5 Ma.

The next diagram shows the likely split up, from 25 Ma, to modern times. Homo is depicted in the bottom left.

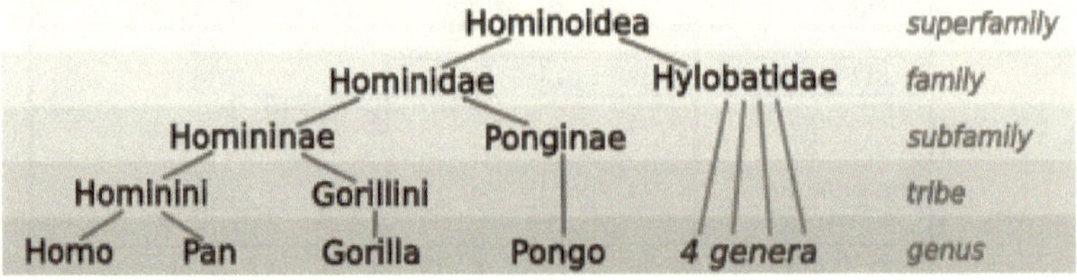


Fig 2.1.5 Simplified sequence of the split of the great ape family, over the last 25 Ma. The final evolution from Africa is shown in the next diagram.

The last diagram shows the emergence of modern Homo sapiens, out of Africa, about 200,000 years ago, as is discussed. The various effects for the rise and emergence of humans are in Part 3.0. There are reasons why humans evolved, and became unique.

The earliest documented representative of the genus *Homo* is *Homo habilis*, which evolved around 2.8 million years ago, and is arguably the earliest species for which there is positive evidence of the use of stone tools. The brains of these early hominins were about the same size as that of a chimpanzee, although it has been suggested that this was when the human SRGAP2 gene doubled, producing a more rapid wiring of the frontal cortex. During the next million years, a rapid encephalization occurred, and with the arrival of *Homo erectus* and *Homo ergaster*, in the fossil record, and cranial capacity had doubled to 850 $cm^3$. (Such an increase in human brain size is equivalent to each generation having 125,000 more neurons than their parents do.) It is believed that *Homo erectus* and *Homo ergaster* were the first to use fire and complex tools. They were the first hominin line to leave Africa, spreading throughout Africa, Asia, and Europe, from 1.8 to 1.3 million years ago.

In this, you might study Fig 2.1.5 and still feel that God played a role, but these effects are buried as fossils. People search for them and dig the fossils up. Other people argue if each of these is a new species, or it is a variation from an existing find.

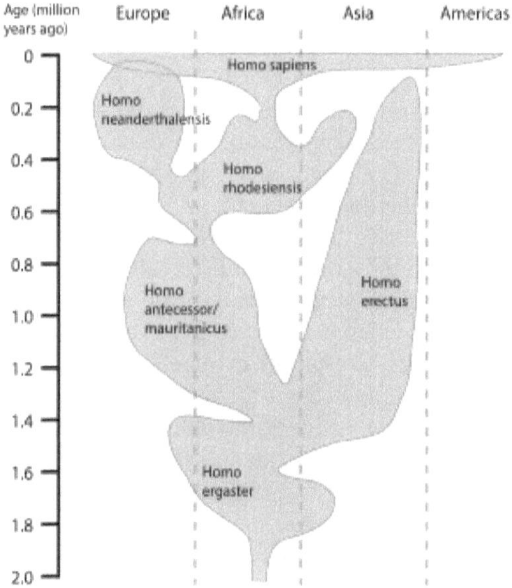


Fig 2.1.6 Diagram of the emergence the human species (via Wikipedia), in the last 2 million years into modern times.

Although these studies, like everyone else, are of the modern rates, the real change was the lead up. Life modified, for complex life leading to modern humans for half a billion years, but it also modified over the huge lead up, almost four billion years since the start. You must think of these huge changes, why it took so long for modern life to come about. Here, my aim in the final parts was to figure how from the half a billion years, we ended with human evolution.

There is also no obvious lead to higher life. Yes, rates occurred, but at each stage organisms seemed to increase brain size, and adaptability. At each stage, life became more flexible. We simply do not know what happens next, but we must account for why life evolved, until a human form was reached. We cannot say, without theory, if these stages of evolution would result in a human form.

At this point, I will stop recounting the fact that life evolved.

Again, this is not the theory. I offer no explanation, of why features died out, or why new species evolved. I do not relate the rise of species to increased oxygen, or allow that maybe it was a different cause. I have no theory why dinosaurs died, only that they did at a time, related to an asteroid strike. I have no theory of why the evolution of life comes in a sequence, such that more advanced forms come after simpler ones. I note how there are ways to test this that take advantage of discoveries in anatomy, or Evo-devo (development biology). Notice, despite facts, there are still differences, as did prelife start 4.1, 3.9 or 3.8 billion years ago, did higher life start at 2.2 or 1.8 billion years ago. Other disciples, such as geology or paleobiology, study these.

The other point of the facts is to reemphasis the huge time for life to evolve. From this, you still might believe in God, but why did evolution take so long? How does God account for the "boring billion", for life to evolve? Without any evidence from another planet, we do not know if the 400 million years of prelife was fast or slow, the 2.5 billion of lower life was fast or slow, or two billion years to the evolution of higher life was fast or slow. It just took time to reach it. Once an advanced stage was reached, the evolution of higher life was in half a billion years, but this was only 15% of the total time that life evolved.

Despite the facts, though, where is the theory?

After reading all of this, you might still believe in God, and that is fine, but there must also be a scientific explanation of how the facts can relate to each other. Preferably, there is a mathematical explanation, how the facts that are observed, can be related by a theory. If someone can take the theory, run it, and reproduce, in mathematics, the explanation of these facts, then we have a full insight.

Let us now discuss the Theory of Evolution.

## *2.2 The Theory of Evolution*

What is the Theory of Evolution?

The previous chapter gave the facts that life evolved. Mostly, these are accepted, other than updates or clarifications. Yet what is the theory of how life evolved, to conform to the facts? Remember too, a fact and a theory is different, especially in science.

A fact works, but it must be rechecked all the time, if new facts are updated, or these supersede existing ones. Evolution has new facts, such as a rise in oxygen levels or types of fossils, as they are discovered. None of the facts though, disproves the basic premise. Facts are imperfect, but over time, they confirm the case.

By contrast, a theory, especially one reducible to an equation, can be perfect, as soon as it is checked. (My theory needs checking.) Einstein's theory of gravitational waves is correct, though written 100 years ago. In evolution, the theory is a statement that if checked, will hold across time, regardless of facts. Anyone should be able to take the theory, and run it. It should uphold evolution, without contradiction.

In the theory though, part of evolution is in trouble.

Darwin's theory of natural selection works fine. I will give it. You can take it, and run it, as often as you want. Whether you are an evolutionist, or a creationist, results will be identical. The trouble is that Darwinism by itself cannot solve everything. One classic example is sex. The theory is that as genes spread fitness rises, but in sex, two parents are required to produce the offspring. It means that cellular selection went down, but the spread of sex increased. No matter how many times that it is tried, the paradox of sex remains unsolved.

Then there is the branch by which life starts. I will show the equation. It seems identical to Darwinian selection, except that the terms that make the equations are different. Darwinian, cellular selection has a term for fitness, and this is for selection in the single species. However, molecular selection has simpler model, without fitness, that works across life. This applies today. The standard view is that cellular selection alone leads to molecular selection, but as fact, these do not tie together. If Darwinian selection often falls, and molecular selection is *neutral*, then this theory cannot explain how evolution changes.

Now, if you tell this to experts, they will laugh.

Years ago, there was a dispute over the punctuated rates of evolution, and other issues. This could be solved if molecular selection was used, but in trying it, the only theory that they had is *cellular* selection. If this is applied to *molecular* selection, it does not work. Experts found that against cellular selection, molecular change can be a gain, but mostly it was *neutral*, or it went down. This caused a dispute. First, it was called 'non-

Darwinian' evolution, but it was then changed to the *neutral* theory. It still is not working, but the *neutral* model is now used as the one.

Nevertheless, if we stop using cellular selection, but find a formula for molecular selection, the theory works. Experts agree that molecular selection began from 4.1 to 3.8 billion years, and admit that it worked by natural selection. Rather, consider a soup where molecules, perhaps of RNA, were competing to spread. Years earlier Richard Dawkins noted it in *The Selfish Gene*. Richard proposed how genes compete to conserve sequence. He gave three examples, but this one was dominant. (Richard used "molecules", not noting if it were RNA, DNA or proteins.)

> "If molecules of type *X* and *Y* last the same length of time and replicate at the same rate, but *Y* makes a mistake on average every tenth replication while *X* makes a mistake only every hundredth replication, *X* will obviously become more numerous."

Richard never developed this,[13] and he assumed that at a point, the gene had to mutate into a new variety, for life to begin.[14]

Instead, let us examine the theory, to see where the trouble is.

Now, in the standard model, genes distribute along, let us call it, a cellular pathway. Of course, life evolves along a *cellular* and a *molecular* pathway, but there is only a model of cellular selection, as a theory. Still, any text will explain how cellular selection works. Take the distribution frequency of any gene in a population to be $x_i$, where $0 \leq x_i \leq 1$. Take the fitness that the gene confers upon the organism as $w_i$, for $0 \leq w_i \leq 1$. Also allow that about the *locus* where $x_i$ acts all the genes competing for that spot have an average mean fitness of $\bar{w}$ (w bar) such that $0 \leq \bar{w} \leq 1$. If the new gene is fitter than average, then $w_i > \bar{w}$. We can denote the change in frequency $x_i$, $x_i'$, $x_i''$ in each generation approximately as follows.

| | |
|---|---|
| First generation | $x_i = x_i$ |
| Second generation | $x_i' = x_i w_i / \bar{w}$ |
| Third generation | $x_i'' = x_i' w_i / \bar{w}$ |

This is converted to a Fisher equation, by the frequency, $\Delta x_i$, as;[15]

$$\begin{aligned} \Delta x_i &= x_i' - x_i \\ &= x_i w_i / w - x_i \\ &= \bar{x}_i (w_i/w - 1) \end{aligned}$$

---

[13] Note Richard suggested two other mechanisms, but this is the dominant one.

[14] Note with respect, I have swapped "X" and "Y" from the original quote, so that it is "X" that copies better and become more numerous. This aligns with my notation later.

[15] It is a minor coincidence, but Fisher's model has been downplayed in favor of the Price model. Note, many models are for the diploid of higher life with sex.

This shows that the fitter $w_i$ makes an individual above $\bar{w}$, then the greater is $w_i$ - $\bar{w}$, so the faster $x_i$ spreads until $\bar{w}$ rises to $w_i$.

Suppose a mosquito population has 1,000 individuals. One individual has an allele, "$x_2$", resistant to DDT ($w_2$ =1) and the other 999 "$x_1$" individuals have a 50% resistance ($w_1$ = 0.5). Each generation with no DDT, ($w_i$ - $\bar{w}$) = 0, so $x_2$ does not increase ($\Delta x_2 = 0$). Yet once DDT is present, $x_1$ halves each generation, while $x_2$ increases from $x_2 = 0.001$ to $x_2 = 1.0$ by about the 18th generation. The curve looks as is next.

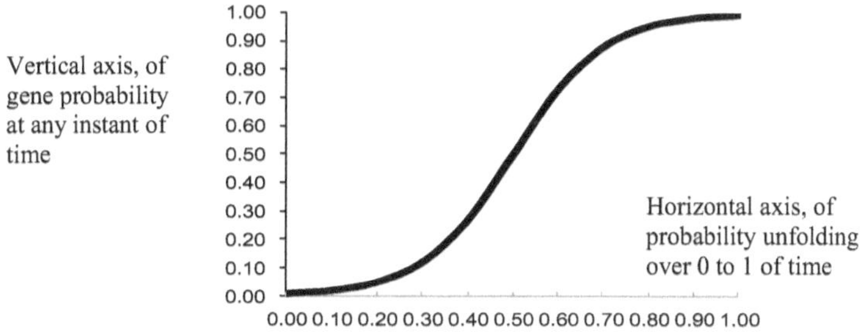


Fig 2.2.1 Natural selection is depicted as a probability, of a single vertical axis on the left. This can be unfolded, from 0 to 1 (or 0 to 100%) as time unfolds. The vertical axis is 0 to 1 in probability. The horizontal axis is 0 to 1 as time unfolds.

Now, although the curve is standard, it does not appear in some texts. For instance, a biology text might explain Darwinian selection, but not show a curve of how it appears. One reason is that modern alleles are diploid, as applies for sexual reproduction. Other reasons might be that no one can combine it with molecular selection.

Nevertheless, assume that the Darwinian curve, of cellular selection, works fine. It is in advanced texts, and you can try this. There is a more advanced curve of selection between diploid species, for sex, but I will not show it here. Rather, assume that the growth of natural selection, as it applies for the growth of an allele, $\Delta x_i$, on a curve, reusing the equation;

$$\Delta x_i = x_i(w_i/\bar{w} - 1) \qquad (1)$$

Assume this curve generally applies for of Darwinian selection. There are variations, but it looks roughly as this.

Now, if you accept Equation (1) for Darwinian selection, the next task is a curve for molecular selection. Again, experts could laugh, but consider this simple model. (Notice, I used a 'k' for selection along the new axis. This was not thought of in my earlier models.)

Suppose that genes (or pre-genes) competed to conserve sequence, so that the more conserved the gene, the wider it spread. Note, molecular losses in early life were high, even to start life, but this is from assuming that genes gained frequency by mutating. However, consider a curve, as a soup of molecules *X* and *Y*, with frequency of $X = X(X + Y)$. Let us replace *X* by $X_k$, for distribution across the *meta*-population of genes. We can make further substitutions. Instead of Y, measure the distribution of genes in which, $\varepsilon_k$ (for epsilon k) is for gene conservation. It is derived as $f(\mu_k, \mu)$, for both $\overline{\mu}_k$ (mutation of the gene), and $\overline{\mu}$ (average mutation). We can then work out a new curve for $X_k$, following the formula; [16]

$$\varepsilon_k = e^{(\log(\mu k) - \log(\overline{\mu}))}$$

This translates into a soup as;

$$X_k = \varepsilon_k/(1 + \varepsilon_k) \qquad (2)$$

Again, I am sorry it looks mathematical, but remember, the statements are based on facts. It was a fact there was a pre-Darwinian selection from early on. It was a fact that it must have gained frequency from early on, regardless of how the modern theory works. The model of selection was from Richard Dawkins, also a fact, which is supported by similar ideas. The mathematics is a theory of the curve. Moreover, the curve works across life, to modern times. It results in a new *neutral* theory, in a way that was not possible before.

Except now, we have a new problem. We can show the *cellular* and *molecular* competition, but not only are the terms different, the formulas are different too. Equation (1) is iteration, which can be stepped on Excel or similar. Equation (2) is a formula, which can be written directly.

$$\Delta x_i = x_i(w_i/\overline{w} - 1) \qquad (1)$$
$$X_k = \varepsilon_k/(1 + \varepsilon_k) \qquad (2)$$

Despite how the equations are different, though, the output is the same. If you traced the output of each curve, it would give the results in Fig 1.2.1 exactly. You could not claim say, that part of Fig 1.2.1 was by one curve, and so the other was different. They are both the same. Especially, one curve is for a fall of fitness over many cases, the other is for a rise. If these curves began with a different formula, but they end up looking the same, there must be a different explanation.

---

[16] This formula works fine for the point I am making. Later, there needs to be a recheck, because genes have many varieties, which do not follow the formula exactly. I am also not sure if the best is log, or alternatively ln, but others can recheck.

Again, evolutionists did not take a curve of prelife, but you can test this, or modify it, if you wish. It uses natural selection. My use of the formula, $\varepsilon_k = e^{(\log(\mu k) - \log(\bar{\mu}))}$, is clever, I hope people try it. It means how $\varepsilon_k$, (that can go to infinity), can relate to the average gene mutation. My use of the '1' in the divisor, relates to the logistic curve. You can try this curve on any sequence of prelife.

The problem though, is what happened when pre-Darwinian selection for prelife, met with Darwinian selection for life. When, as I call them, molecular selection met with cellular selection.

According to experts, molecular selection must cease, to be replaced by cellular selection, or the two will not work. Yet how did the change occur? You could not have 500 to 100 million years, molecular selection, without fitness, across life, and absorbing heat, suddenly changing, at an instant, to cellular selection, with fitness, per species, and now radiating heat. There is no theory of how such a change occurred. Besides, if you do not allow continuity, we are left with the paradoxes. It applies for early selection, for applications such as the evolution of the chromosome, for sex, and for modern molecular theory.

This is where evolution becomes complicated, so let me introduce a strange theory to solve it.

First, you are free to dismiss my theory, but if you do, how do you solve the paradoxes. Experts can write books, say, about sexual selection, but still it is not solved. There is also no theory of molecular selection, or an explanation of the *neutral* rate. There is no theory of the first 500 to 100 million years of evolution, despite how experts admit that it was by natural selection. The existing theory in an equation will not add up. Somehow, an expert, a mathematician, has warned other experts that if you write the equation of pre-Darwinian selection, the addition of the two components will not work. To write a book you must first choose which one. Even if the two curves do not follow my curve exactly, a fundamental effect is not working.

Notice too, no terms combine equation $x_i$, $\overline{w}$ and $w_i$ in a Fisher model with the equation of $X_k = \varepsilon_k/\sqrt{(1+ \varepsilon_k)}$.[17] However, I have tried to combine these on earlier sheets. The new term appears to be like;

$$z_{(i,\,k)}(t) = (x_i\,(t) + j\varepsilon_k)/(1 + \varepsilon_k) \text{ where } j = \sqrt{-1}$$

The first 500 to 100 million years selection was in real time, but at 3.7 to 3.5 billion years it changed to Darwinian selection. I used $z_{(i,\,k)}$ component for a combined curve, where 'j' is used. Notice, if the 'k' effect is not there, then $z_i = x_i$, across the left vertical axis. The molecular selection uses a 'j', to prove it is on an imaginary axis.

[17] Notice, a gene must transit an 'imaginary' valley, via a 'real' horizontal transfer.

The solution is that molecular selection, the first, becomes imaginary. This keeps molecular selection until the present day, so the paradoxes of evolution, where genes keep gaining frequency, are now solvable. The problem is that you cannot have two forms of selection, each summating to 100% gain, on one pathway. There has to be a change, not in facts, but in how humans do the mathematics. Take Darwinian, cellular selection, as real. To stop molecular selection changing in the existing theory, take molecular selection as imaginary.

Fig 2.2.2 shows how the theories add. There are two logistic curves, the front is Darwinian, *cellular* selection; the back is for pre-Darwinian, *molecular* selection. The curves join on a left vertical axis. Here, the gene gains mutation along a front cellular pathway in real time, as shown in Fig 2.2.1. However, at the same instant, any gene also gains molecular selection over time, along the 'k' axis. For instance, certain genes gain the far gray line, as they spread in all populations, as shown on the diagram. Notice, although the curves are logistic, the front curve ends in a single point for gain in one species. The back curve, for molecular distribution, ends in a line that represents distribution across all of the populations.

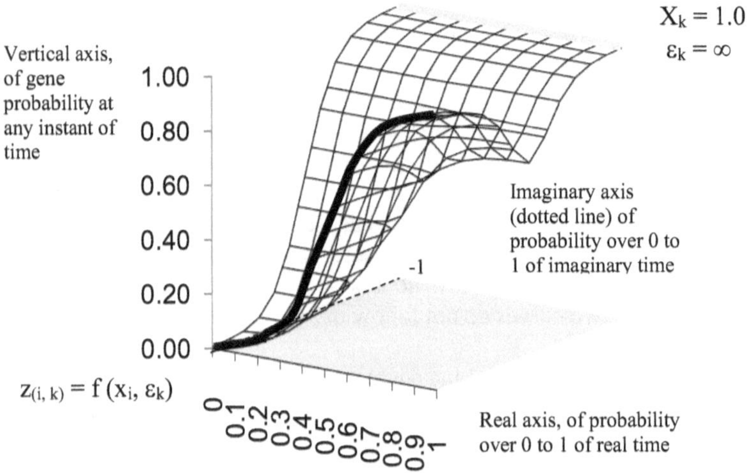


Fig 2.2.2 This shows a gene also gaining along a single left vertical axis, but the axis is unfolded along two pathways of time. The front axis show 'real' gene spread, along a real time. The rear axis shows spread along a new time, as the gene spreads across many populations. Genes, such as ubiquitin or histones, or hox genes in animals, trend to the furthest spread along the far gray line.

Suppose that experts accept the curve (allowing for modification), but they insist that if cellular selection can be proven by one gene, there must also be one gene for cellular and molecular selection together, if it were one gene. Again, the math works. Even for one gene, the mathematics of molecular selection will still gain at a 90⁰ "lead" over gain for mutation. Both schemes still add, and the rules of probability are not violated, for the two-pathway solution.[18]

Of course, experts also might be startled at Fig 2.2.2.

They can argue that there is no selection for molecular evolution, but in prelife times, selection is in real time, but pre-Darwinian. The problem is when Darwinian selection begins, to combine both in one diagram. (This is historical.) I have chosen to start the Darwinian selection, about 3.5 billion years ago, but allow that prelife started about 0.5 billion years before. By modern times, the figures match again.

Now consider the problem of sex. This is a problem in the standard theory, especially as sex falls for the *neutral* rate. The next curve shows the curve modified to allow sex. This often occurs at $\varepsilon_k = 1$and $X_k = 0.5$ for a *neutral* rate of 6 x $10^{-9}$ mutations per gene. Please test this, but the figures are historical, and they do match.

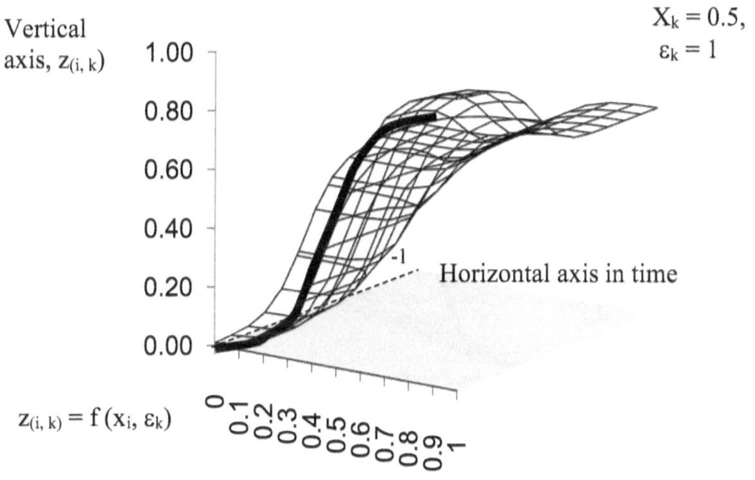


Fig 2.2.3 The curve before redrawn for sex. The front curve (the heavy black line) is the same. However, the rear curve has been drawn to show the effect of sex reducing to the neutral rate. The curve is fully explained in Chapter 2.5

Now, the curve for sex is for when the far line falls, down often to the *neutral* line. (The grey line is not shown.) When I first drew this, it was for

---

[18] It needs to be worked, but the gain for the front axis slows down over the standard, as the gene becomes more conserved. In higher genes the gains are more along the neutral rate, shown later in Fig 1.5.1.

the earlier figure only. I was pleased that the new model justified, say, the '1' divisor used earlier, and it showed how the curve settled, for the evolution of sex at lower figures. I will explain this more in Chapter 2.5, on the Paradox of Sex.

However, having drawn for sex, at the *neutral* point, I was curious how it will work if $\varepsilon_k = 0$ and $X_k = 0$. I had expected the grey line to drop to zero, but was surprised (initially) to discover the curve looks as shown in Fig 1.2.4 next. You must remember (we all must) that the diagrams are not a 3D landscape, but a single probability in two lots of time. I first thought that the far line would drop to 0, but it does not. When you feed in the numbers, instead of falling, the front curve is repeated across all the scenarios. This strange case validates the standard theory. It means that if $\varepsilon_k = 0$ and $X_k = 0$, that the standard front curve (black) will apply equally for all cases. The rule is that for fast mutating genes, only the $x_i$ (as $z_{i,\ k}$) alters, for anywhere below the *neutral* rate.

The diagram for $\varepsilon_k = 0$ and $X_k = 0$ is shown. Except that, for early life, sex, and the *neutral* rate, the gene will accumulate over time. This is already proven on the first two curves, Fig 2.2.2 and Fig 2.2.3

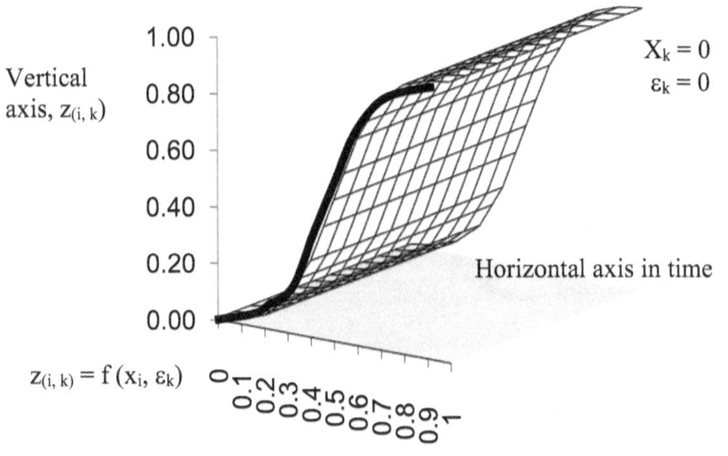


Fig 2.2.4 Previous curve redrawn to show $\varepsilon_k = 0$ and $X_k = 0$. You might think for this case, the back plane would lower to 0. In fact, the numbers just repeat the front curve, where the standard case holds all the way through.

Still, there are problems. There is a gap between prelife, as Fig 2.2.1, and the evolution of sex, from (1.2 billion, see facts) through to modern times. The there is a temptation to assume that the evolution of sex must have caused a change in the basic premise. Thinking about this, I am convinced now that his did not occur, and the basic premise continued in a modified form from the start, at 2.2 billion years when higher life first evolved. There is yet no theory of how genes spread across lower life. We can only note that the first chromosomes were circular. Horizontal transfer was also more frequent in earlier organisms. As well, my equations emulate a continuous increase in conservation for a gain in distribution. However, the increases in conservation also evolved in "steps". For instance, the change from low to higher life included a major step. These major step increases are not in standard theory, but they are not here either, apart from noting that these steps occurred.

At this point, I will stop the Theory of Evolution.

If you wished to know how evolution worked, the gross theory, the previous curves show how. The first curve, Fig 2.2.1, shows Darwinian selection, how it is described. You can copy this in all its other forms, but the theory is correct, as per the textbooks.

However, this theory of Darwinism cannot reveal the full effect.

For instance, it is fact, genes also spread across species. Ubiquitin, or the H4 gene in histones, are spread across all of higher life. How do you explain this result, by a theory that works by Darwinism alone? There is the theory of prelife. Experts admit that it existed, 500 to 100 million years, and it was by natural selection. Except, prelife, did not contain fitness, and it must have existed across all species. How can you explain this, if you are not prepared to write the equation?

Again, this is a theory that anyone is welcome to test. All the books, with due respect, that explain how evolution works as description, can easily be worked as the theory. If it reproduces the curve in Fig 2.2.1, or a similar type curve, then it is correct, based on math. The problem is that there is no extra curve, of how molecular selection works, from prelife, and then down through the ages to modern times.

The next few chapters will explain how the facts and the theory can come together, to prove how evolution works.

## 2.3 How Did Life Begin

How did life begin, on this Earth?

Well, before we start, two adjustments must be made. The first theory is my own, how molecular selection might have absorbed heat. Experts think that the first selection was Darwinian, but with no model of how it acted. By contrast, I use the Richard Dawkins theory that first selection was to conserve sequence. However, it acted on an imaginary axis so it could work on early life. Suppose the first traceable molecule was the start of 23s RNA. This would apply across first life and it would work by molecular selection acting to form it.

Fig 2.3.3 shows a newer version from the evolution of archaea and bacteria. The lower version of LUCA is not correct (there was no LUCA) but the later developments seem correct. Again, my theory teaches that prior to archaea and bacteria there were no "first cell". Instead, molecular selection worked on pre-cellular life. It was only once the pre-cells had formed, that archaea and bacteria evolved.

Another point is the early evolution of the ATP molecule. The next diagram shows how it works, but this is already complex. Notice, in this the evolution of ATP is complex. My theory states that the first molecules were formed by molecular selection, but already, even this early diagram is beyond that. The lipid layer, say, shows that cells were already forming, even in pre-life. Rather, even ATP must have gone through several steps to form. Wikipedia details how.

Now, let me quickly recap.

First, "my theory" is that the evolution of sex is a result of selection having to modify along real and imaginary axis, and I will stick to this. However, Nick Lane and others have suggested a different problem, that the conflict was the evolution of mitochondria, essential to the evolution of eukarya. To illustrate this, they have shown the evolution of early life, more detailed than I (that I include here). Until now, I held the standard version of early life that it occurred as per Fig 2.3.1. However, the split of archaea and bacteria is now detailed, and is to be used instead. The split for the ATP module is also given.

In my model, the theory follows a logistic curve, and moves from 0% to 100%, if it is successful. Except, cellular selection also moves from 0% to 100%, but the curves are not the same. No one will admit this, but it is where the standard theory strikes a problem. Anyone could formulate a theory for molecular selection, but if you add it to *cellular* selection, this becomes disjointed. This is where results do not add. My solution is that in the theory, the first, molecular selection adds along an imaginary axis. (I claim that the universe works a similar way, but critics can object that I cannot do this, because it is so unfamiliar.) Rather, if you allow that life began with molecular selection, but it then leads to cellular selection. We can call these two axes fast and slow, but there a formal term. The cellular axis is in 'real' time, but the axis along which genes gain frequency is 'imaginary' time. There would be proteins, first 16 of them, but later it was 20. The walls of archaea and bacteria would have begun, but these were not directly connected. There were primitive cells, but no selection of the advanced cells. Prelife might have absorbed heat. We cannot define this as "life" (as understood) until it could radiate heat.

Instead, the start of selection at near 3.9 (or 4.1, or 3.8) billion years ago is a fact. Evolutionists also admit that when it started, it was natural selection, but that it must be "Darwinian". There are such models, but none of them works. Evolutionists admit that selection must have had similar results for the first organisms, yet no one can put it together. But another model is possible. Richard Dawkins had one model, and it is easy to prove that it used natural selection. Rather, the issue is when pre-Darwinian prelife combines with life, as Darwinian selection. In the expert view, the types of selection cannot combine. One must cease, and be replaced by the other. The proposal by experts is the axis of cellular selection replaces molecular selection, but it will not work. When prelife model and pre-Darwinian selection met with Darwinian selection, there is no answer.

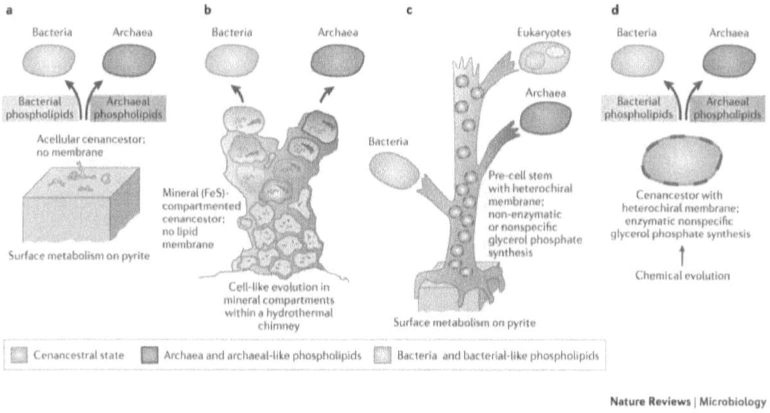


Fig 2.3.1 The diagram shows bacteria and archaea starting from a similar point. This diagram, though crucial, is hard to find among the others, which more show the three trees of life originating at a common point.

Instead, my theory follows the facts. If anything, we can move the problem of first life back.

Before there can be life as we define it, RNA, DNA, ATP, and early proteins must evolve. Early prelife is also complex. Consider the 23s RNA. It had to convert an RNA replicate into proteins; it is a large molecule, which evolved in six stages, or more. It evolved subunits and it was circular. There must be an early, thermodynamic process, and this favors early molecules that form in circular formations. There is also a case that the first molecules were based on RNA, which was first, and is simpler in structure than DNA, which evolved later.

This is from a hypothesis that life began with an RNA and Ribosome world, before transferring to a more modern, RNA, DNA, Ribosome and Protein world. The RNA scenario is also shown in Fig 2.3.1. To explore this, a standard view is that mutation and conservation are opposite. Suppose a gene gained 100% by mutation, and a 100% by conservation. In a conventional view, the two would cancel. Instead, if it were at $90^{O}$ instead of $180^{O}$, then the two would add. In this, gain by conservation is a $90^{O}$ lead over gain by mutation. If we added these gain would be mutation (at $0^{O}$) plus conservation (at $90^{O}$ lead). The result would be a net gain, by a combined line, shown in the diagram next.

Here, molecules were first trying to copy accurately. There was no fitness of cells competing, which did not exist, and early molecules might have absorbed heat, rather than radiated it.

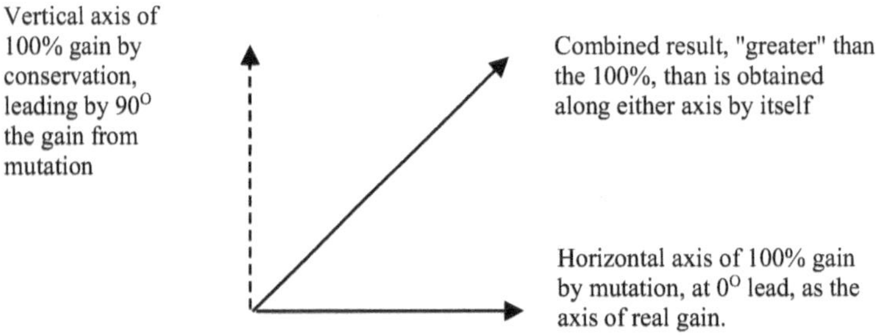


Fig 2.3.2 When two components of gene gain, one for mutation and the other for conservation, the result is a net gain where both go up. To those unfamiliar it seems strange, but if genes gain in both, as life has evolved, the two will add.

All we know is that in an early stage, molecules, which can replicate accurately, gained an advantage over those that did not. There were also pre-cells as an outer coating that formed, to assist replication. In this, the first replication of RNA is not so bad.[19] John Sutherland and others have found that the uracil and cytosine bases could form, but be assisted by intermediates for the adenine and guanine bases. For 500 to 100 million years, before life, there can be intermediates. Prior to proteins, structures

[19] Jack Szostak and others made the point, that these linkages favor increased replication, for the generation of pre-cells. Jack Szostak showed how various combinations were tried. Remember, if the goal were longer RNA and DNA, conservation for sequence would help.

were derivations of ribosomes. It does not require pure RNA, and there might have been early DNA, or other ribosomes.

If anything, if we take the first replication back to the start of prelife, there are many possibilities. Again, there are difficulties modifying pure RNA, so maybe it began as mixtures of RNA, DNA, or other molecules. There have also been RNA base structures seen in interstellar objects. Perhaps part of the early formations was to use these, in early Earth. Remember, the first goal is to grow molecules longer, and it evolved over hundreds of millions of years. Perhaps at 3.9 billion years molecules began to replicate, but then millions of years later, RNA and DNA were achieved. That replication was for conservation, and RNA chains can retain conservation for billions of years.

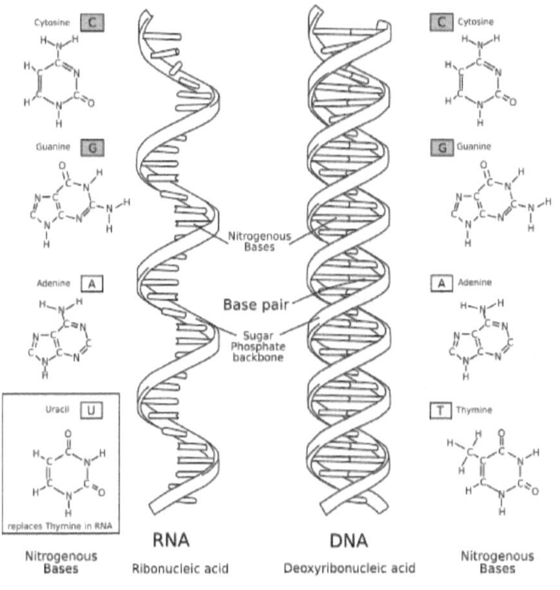


Fig 2.3.3 A copy (via Wikipedia) of the RNA worldview. Everything said here, genes first competing to conserve sequence, and the evolution of complexity, from a simple beginning to develop mutation later, supports this view.

The other problem is the motive of life to evolve.

Again, the modern molecular, neutral theory, teaches that life is either *neutral* or a loss of fitness, but it cannot be correct. In pre-life, the motive was to conserve copy, so this must have been a gain of fitness. However, there are other problems. One is that it benefits molecules to evolve into a circle first. It changes, but RNA or DNA formed circular molecules. It also benefits early molecules to form circular structures, such as for the 23s

Ribosome. Again, I have no direct proof why it is this way, but when life evolved, molecules would form in circular structures, just as it provided pure forms to replicate, rather than mixtures.

However, if the first genes tended to RNA, these would fold into 3-dimensional structures, as a start for the 23s Ribosome, and other RNA structures. To place genes in a circle would need DNA, which came from RNA. It is not certain if early proteins formed themselves, or resulted from the 23s Ribosome folding. Instead, by the end of pre-Darwinian evolution, cells contained RNA, DNA, ATP, ribosomes, enzymes, and proteins. The pre-cells also contained vesicles made of ether linkages for pre-Archaea and ester linkages for pre-bacteria.

Even so, just to form the RNA world has problems. RNA is unstable by itself, and is soluble in water, and so on. There are other objections, such as the early chemicals, that produce the uracil and cytosine bases prevent the adenine and guanine bases from forming. However, we must remember the power of natural selection (although, strangely, I seem to against the evolutionist). Evolutionists want to see "life" begun as Darwinian selection from a pre-biotic soup, but it did not happen that way. For longer than the dinosaurs evolved, evolution was natural selection, but a pre-Darwinian form. There was a first stage of evolution, but it was not "life" as defined in the current texts.

Again, the start of the tree was the evolution of pre-archaea and pre-bacteria, with these genes as common. There is no evolution of archaea, bacteria, and eukarya, starting at once. I suggest a new tree be drawn, showing eukarya coming after archaea and bacteria have evolved. The tree for eukarya should be redrawn on top, with roots in both archaea and bacteria. The branching of mitosis mutating into meiosis was overlooked as well, in the evolution of sex.

On the other hand, certain genes came after Darwinism began. For example, the flagella of archaea and bacteria, is quite different, so these must have evolved separately. My opinion (now confirmed) is that the split between archaea and bacteria was early, almost pre-Darwinian, with core genes forming in the pre-branch.

Again, the aim is to prove Darwinism, but it cannot have begun pre-3.7 to 3.5 billion years, when cells or competition between individuals was not there. Pre-Darwinian selection was for conservation. If copying of RNA, it is not accurate, and other factors and molecules would be mixed in. If the 5' to 3' copy of RNA were the most accurate, this would gain advantage over other types. Textbooks assume that life began with Darwinian selection, but if you compare Equations (1) and (2), outputs seem the same, but the backgrounds are different. Equation (1) is a fall of fitness, but Equation (2) is a rise. Equation (1) does not replace Equation (2), but it adds on to it.

In summary, this chapter is crucial to my theory. Anyone can object that there is no need for molecular selection, but how can we account for the first 500 to 100 million years. It occurred, it is in the textbooks, but there is no theory of how it worked.

Rather, I challenge evolutionists on how they will solve it.

If natural selection had occurred in pre-life, even if pre-Darwinian, it must have had an effect that can be modeled. It might have absorbed heat, it could have been across all of pre-life, but the effect was there, for 500 to 100 million years. I have a model that works. Evolutionists can use this, or another model, but it explains prelife. There were pre-cells of ether linkages pre-archaea and ester linkages pre-bacteria. The pre-cells enclosed genes and molecules. There were ribosomes, such as the 23s, and other RNA structures, such as elongation factors. There were rings of DNA enclosed in the cells. There was ATP enclosed.

Once replication of the pre-cells had gone as far as it could, over 500 to 100 million years, a new type of evolution was needed. Nature had to fully self-replicate, and the pre-cells had to change from an absorber of heat, to self-sustaining ones that could radiate heat. Pre-Darwinian, *molecular* selection, could no longer functions by itself. A new type, of Darwinian, *cellular* selection, was required to be added.

Full, Darwinian evolution, could start from here.

## *2.4 The Flowering of Darwinian Selection*

At 3.7 to 3.5 billion years ago, the Darwinian flowering began. The 500 to 100 million years of pre-Darwinian evolution could go no further. Now, a major change was needed.

All these statements are true, based on facts. There were 500 to 100 million years of pre-Darwinian selection. There was Darwinian flowering at 3.7 to 3.5 billion years ago. Change was also needed. The problem though, is not the facts of the statement, but my theory of how the change occurred. To experts, the Darwinian selection replaced the pre-forms. In my view, the Darwinian change added to the existing selection, but never replaced it. If you consider all disputes in evolution, even the evolution-creation controversy, it seems that some of them are more important. Yet no, of all the other disputes, even the evolution-creation controversy, the truth is, no one cares. However, if Darwinian selection added onto, rather than replaced the pre-forms, my theory contradicts the experts. Only one view can be correct.

Please consider again the expert view.

The way it is depicted in the standard texts it cannot work. Selection could not go on, for 500 to 100 million years, with a gain of fitness, but absorbing heat, and then cease. Especially, if early Darwinism was a few features, but the remainder stayed as pre-Darwinian selection. After all, even in the modern theory, Darwinism, as *cellular* selection mostly is neutral, or a loss of fitness. These had to keep gaining pre-Darwinian, whereas changes of full replication, or the radiation of heat, occurred under a Darwinian influence.

If anything, Darwinism bought several changes. There had to be full replications, of both the genes and the outer cells. The cells now had to absorb first, and then radiate heat, so they could move to colder places. Genes also had to develop their own expression, unique to the species, by competition with rival species. There were about 500 genes in archaea and bacteria, but genes never stopped competing for conservation, from prelife to the present.[20] Despite mutation, short RNA lengths billions of years old, are conserved today. When eukarya seemed to evolve, 2.2 billion years ago. Instead of falling, conservation of gene sequence increased $10^{-7}$ to $10^{-9}$. A histone H4 is 98% conserved near its 300 bp length, not possible in pre-eukarya. In eukarya, not the mutable, but conserved genes that are the widest spread.

Another problem is the mechanisms of the change. Much has been written about underwater vents, or changes of hot to cold. The problem again is that for evolution of archaea and bacteria, this might have been in two places or at different times. It is unlikely one evolved and changed,

[20] Another quoted fact, but it may have been less for first archaea and bacteria.

for instance, archaea into bacteria. New suggestions (Nick Lane and others) are that archaea and bacteria evolved simultaneously, but then separated as the cells competed.

To show the difficulty, the next, top leftmost corner shows proteins in eukarya, in a list, from most conserved to least conserved. (I was not able to fill it in, but it can be checked, and filled in.) The first column shows protein changes. Then it gives nonsynonymous DNA (that affect fitness) changes, then synonymous DNA changes (with no effect on the fitness). Molecular evolution has to follow the left column, of nonsynonymous change, or the rightmost column. Experts state that it is the rightmost column, and when they view this table, they think that I made a mistake.

| **Protein Frame** | **Protein Change** | **Nonsynon. DNA** | **Synonymous DNA** |
|---|---|---|---|
| Ubiquitin | 0.000 | $>10^{-14}$ | Near Nil |
| Histone H4 | 0.005 | $>10^{-13}$ | $6.12 \times 10^{-9}$ |
| Histone H3 | 0.007 | $>10^{-12}$ | $6.38 \times 10^{-9}$ |
| Aldolase A | | $7 \times 10^{-11}$ | $3.59 \times 10^{-9}$ |
| Cytochrome C | 0.110 | | $\sim 6 \times 10^{-9}$ |
| Hemoglobin | 0.600 | | $\sim 6 \times 10^{-9}$ |
| Serum albumin | 0.950 | $9.1 \times 10^{-10}$ | $6.63 \times 10^{-9}$ |
| Interferon-β1 | | $2.2 \times 10^{-9}$ | $5.88 \times 10^{-9}$ |
| K Casein | 1.650 | | $\sim 6 \times 10^{-9}$ |

Table 2.4.1

Rather, exerts claim this is a listing, of the most conserved genes down. Explanations are that change represents a resistance to mutation. One biologist argues it is "hard put to argue that histones were more important than immunoglobulin's",[21] ignoring that histones are 100 preserved in eukarya, but immunoglobulin is confined to animals.

If anything, if Table 2.4.1 included virus genes, there are *billion*-fold difference between mutation of genes for ubiquitin and virus genes, ($10^{-13}$ to $10^{-4}$). A *billion*-fold difference in any science is huge, but in biology, there is no theory of how the differences relate.

Rather, experts focus on the synonymous DNA rates, because they can relate to the *neutral* theory. (Actually, the rate is combined, but the further down, the longer the synonymous rate becomes, so rates become more similar.) This theory solves nothing about paradoxes, but it does tie to a molecular clock. People ignore that my theory too has a *neutral* rate, with

[21] By John H Gillespie, this seemed to admit that he could not see a correlation.

predictions, such as genes that mutate slower than neutral still gain via sex, but those that mutate faster lose via it.

Moreover, my theory has the top, leftmost genes as most conserved. Ubiquitin, H4, and part of H3 are 100% conserved for eukarya. Further down are less conserved genes. Albumin is in egg laying species, and interferon is in higher animals. K Casein is in milk producing bovines. This predicts not only a *neutral* rate, but from the leftmost columns, the rate at which genes spread, across all of life.

If we now think that genes gain conservation as life evolves, we can draw a scale. Fig 2.4.1 shows an approximate sequence of mutation.

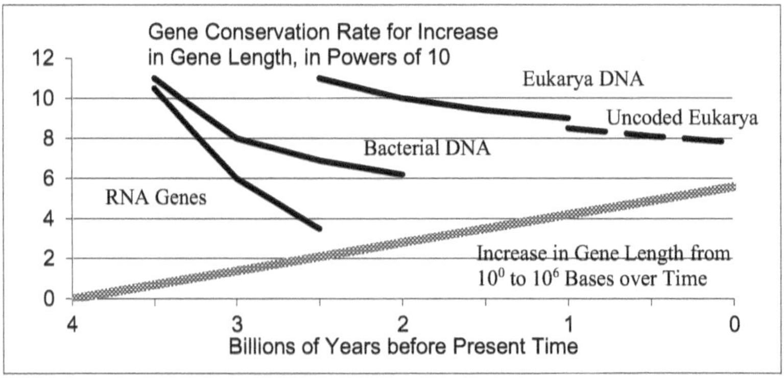


Fig 2.4.1 Increase of conservation rate (or *fall* of mutation) for a steady increase in gene length over time, if average conservation undergoes "step" increases. (Note, I have left the first 400 million years blank, until I can fill it in.)

Assume that RNA was first, from a short lengths then growing longer. This followed by DNA genes falling to $10^{-7}$. Eukarya began high, but fell later, with a higher rate for uncoded eukarya. Of course, this drawing is an approximation. More likely, early genes were highly conserved for lengths of about 4 codons, or 12 RNA/DNA codes. Later, gene length increased for DNA, so conserved sequences could then reach near 100 codons, or 300 DNA codes.

The other trend is that as a gene grows longer within a domain, gene conservation falls with increased length. If DNA codes in eukarya are about $10^{-14}$, but the gene is $10^{3}$ long, rate will fall from $10^{-14}$ down to around $10^{-11}$. Eukarya contain extra, uncoded DNA, which falls faster still, as uncoded lengths grow. This is why H4 is fixed at 300 bp length. However, additional histones (H1, H3, etc.), have longer code, and more uncoded sequences, so mutation rate falls in proportion. (Note, the H4 was first, but others followed.) Fig 2.4.1 shows a 'rising' (barely visible) gray line for an increase in gene length.

By contrast, Fig 2.4.1 shows how it might have developed.

Early selection for molecular evolution did not cease once cellular evolution had begun, but continued, and cellular selection added on its development. Of course, it seems odd if a gene can conserve sequence, but also mutate at the same time. However, there are two answers. One is that the genes doubled, such as RNA to DNA. The other is that while it is not admitted, the chromosome is a paradox. If two genes compete, one for conservation, and one for mutation, they compete at $90^{O}$ (not $180^{O}$) so there is mutual advantage where both gains.

Especially, far back, where first genes were of RNA, and cells were forming, differences were pronounced. Cells reproduce by doubling the entire organism, by splitting material in the original cell. By contrast, molecules (DNA is shown, but the first to replicate was RNA) copies the sequence exactly, with new materials added for each branch. (RNA must replicate as one. This was is another reason, when back in early life RNA was replaced by DNA.) Fig 2.4.2 shows the differences.[22]

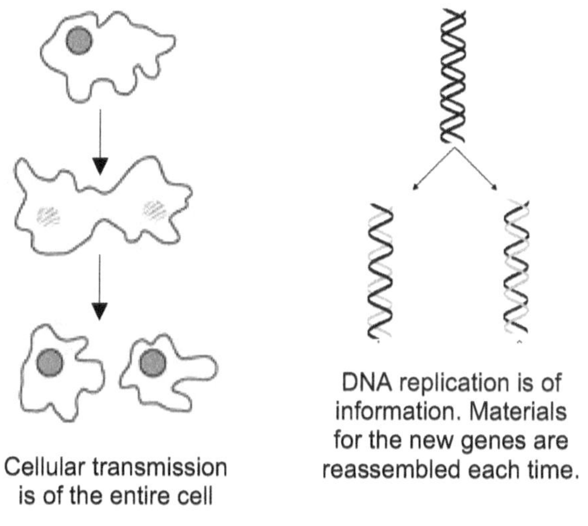

Cellular transmission
is of the entire cell
DNA replication is of
information. Materials
for the new genes are
reassembled each time.


Fig 2.4.2 In cellular reproduction, new materials are absorbed into the parent cell, which passes into the next generation. If the cell dies, or fails to grow, it cannot pass on offspring. By contrast, in a gene does not live, grow, or die. A gene replicates its information, which is reassembled into a new gene on the spot.

Moreover, the cell is a native of the body parts that make it up, so that one cell must exist to reproduce another. (Again, the first pre-cell led to archaea

[22] Fig 1.4.2 shows how DNA replication is 'semi-conserved'. In late prelife, DNA came to replace RNA, but retained the simple basis of RNA replication.

and bacteria.) In contrast, as long as the RNA/DNA sequence is known, similar molecules from a source can repeat it.

Now, consider the evolution of the first chromosome. Again, it seems incredible, but for all the discussion, there is no frequency model of why the chromosome evolved. In a standard model, a gene in the first species gains 100%, but there is no reason why the gene joins the chromosome after that, because it is at 100% already. Of course, books are written with complex mathematics, but none can prove why for a single pathway, there is an advantage to the chromosome.

To see the problem, suppose that life required, say, the elongation factor to start, common in pre-archaea and pre-bacteria. Call this gene AAA. However, life also needs another gene, say, to provide mobility for the different cells. Call these genes as X and Y, to enable life. Two forms of life are possible. Gene AAA, key to life, will be in all branches, but X or Y can be added to either branch, for life to start, as AAA-X or AAA-Y. Notice, gene AAA is 100% in both populations, but genes X and Y are 100%, in each branch. Genes X and Y cannot produce life without joining AAA, but once each does, it gains by joining the chromosome. From this, we can extend the model further.

Again though, it is not that simple. No pre-eukarya gene was 100% conserved, but they must have started that way before the archaea and bacteria split. Once the split occurred, the nonsynonymous genes did not vary more than 50%, but there were small changes. If we go beyond the initial split, then there are changes as new chromosome evolved in each line. However, it is only once we move to eukarya, we can see near 100% conservation in ubiquitin or histones, or the hox gene in animals, but with extra chromosomes added in various lines.

The next problem is that once there are two-axes of gene spread, and then genes can compete both 'across' and 'along' the chromosome.

It is harder to see in lower life where chromosomes were circular, and genes were embedded as semi-chromosomes. However, in eukarya, the layout of the double chromosome makes the issue clearer. Remember, eukarya genes are in double layers, and we differentiate the genes that evolved first. Genes then compete 'across' organisms, so that one group can gain frequency over the other. This is shown by a solid line in Fig 1.4.3 (next page). However, in two-axis theory, genes compete 'along' the gene grouping. For example, a slight change of frequency in favor of one allele, leads to an adjustment for a new species. Still, there is no way to prove how higher life evolved from lower life.

For the solution, we need to look at how higher cells have organized. If there are two organisms, genes compete across the organisms as shown by the solid line, for standard evolution. However, genes also compete along the organism, as indicated by the dotted line shown next. Fig 2.4.3 shows groupings, of genes in higher life, of genes doubling, and double-genes forming groups. People can object that in standard theory it can be a single gene, competing against, itself. Well, the same math holds for a two-axis theory. If we allow that conservation is on an extra axis, it leads cellular selection by $90^{O}$. No one needs comment, for evolution as the gain of a gene, but the one or two-axis model works. Genes can gain along either axis, in the one or two axis theory.[23]

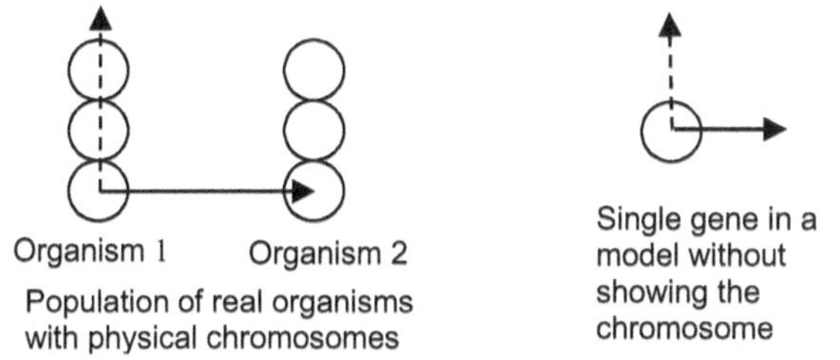


Fig 2.4.3 In the standard frequency model, genes only compete with other genes in rival individuals 'across' the chromosome (solid arrow) for *mutation*. However, genes also compete 'along' the chromosome (dotted arrow) for *conservation*. In a combined model, a gene can compete "with itself", along both axes together.

Still, to evolutionists, Fig 2.4.3 does not seem possible.

Theorists have rejected the concept of molecular selection, and this is not just political. There was a dispute over the paradoxes, that could be resolved if genes selected for molecular gain.[24] However, it was testing it against cellular gain that it went awry. If the gene gains by mutation for cellular gain, to prove a gain by conservation gives a failed result. Instead, for a gene to ensure its lineage, it has to copy its code exactly, so genes can be reassembled from new materials each time. Gain from mutation uses a term for fitness, but for gain by conservation, the fitness is not needed, and a simpler replication is used. This lacked Darwinian fitness, but it could not continue for that time, but then vanish. Darwinian selection was only starting. Even in modern theory, we have paradoxes of how the types of selection add together.

Rather, early selection kept going, but it moved to the background over time. Richard Dawkins, for instance, stated that there is nowhere that replication for prelife changed life, but it was without considering that

---

[23] Centuries ago, mathematicians did not know what an 'imaginary' number was, so they decided to make it act at $90^{O}$ to a 'real' number. It is interesting then, that a competition in the chromosome that results in an 'imaginary' number in an equation, also enacts at $90^{O}$ physically, for competition 'across' and 'along ' the locus.

[24] This is a reference to Shannon's famous theory on the quality of information. A code first has to replicate information accurately. Mutation, seen by many as a key role for the gene is an increase in disorder, so this point too, is misunderstood.

*molecular* selection was first. It was only later did this join together with *cellular* evolution. If we represent cellular evolution as when the gene mutates, it is where replication for conservation at $90^{O}$ lead is joined by mutation, at $0^{O}$ angle. This is where mathematically prelife evolves into life. Prelife would begin as molecular conservation, but cellular selection now occurs along the real axis of time, and molecular selection becomes imaginary.

Moreover, no existing equation can prove how nonlife evolved into life, regardless of mathematics. Equations for a gain of gene frequency are *reversible* only. There is an *irreversible* equation, but this only proves why thermodynamics is within the constraints of the Second Law. The problem is that life in a restricted sense seems to violate this law, in that life generates a net gain of order. My own equations do not solve this. However, a rule of molecular selection 'leading' cellular selection can set a direction for the evolution of life, greater than possible from existing reversible equations alone

Not only in evolution, but also in physics, astronomy, or quantum theory, no equation can prove why a *reversible* process proves that life evolves in a particular direction. *Irreversible* equations are for a decrease in order, but the universe could evolve forward or backwards in time. However, if molecular selection leads cellular selection by $90^{O}$, it gives a direction to life, provable by *reversible* equations. The direction in which the universe tends towards disorder, provable by equations, is also the direction in which life evolves, provable by equations.

Genes reproduce physically, inside organisms, and they pass on to offspring physically, like passing a baton in a relay. Yet genes only reproduce information. In the famous polymerase chain reaction (PCR), humans provide chemical ingredients. It is information in the DNA, not the chemicals, that is multiplied millions of times. Organisms play two roles when transmitting DNA. By reproduction, organisms are a chemical relay station. By selection, organisms are a way to modify DNA information. DNA as molecules is copied as a process, but change of sequence occurs physically, even in the past.

If we move beyond first life, we come to other paradoxes, such as the evolution of the first chromosome, or the evolution of separate species. Again, there are descriptions, of one set leading to another, but the math of how it works is not solved. Beyond first life, the next paradox is that of sex. Hundreds of book and thousands of papers have been written, and people are convinced that this is solved. Here, the facts prove that higher life evolved from 2.2 billion years, but perhaps it did not complete until 1.2 billion years, when sexual division was proven. The theories are that life split into three, of archaea, bacteria, and eukarya. The split into three is correct, but the sequence of how three branches originated from an 'un-routed' branch is not. Again, the increases would have resulted from both conservation and mutation. Longer proteins, for higher life had to evolve, and many factors cause them. The problem is the mathematics. This registers changes by mutation, but not a gain across life. If we can solve this first, then the rest of the changes fall into line.

If anything, there might also be an advantage to thermodynamic direction. The early world was hotter than now, and we think of life as radiating heat. However, the early molecules could also be absorbing heat. The equations for frequency do not say about temperature, which can go either way. Perhaps thermodynamics began in a hot environment, with the vesicle in which life began slightly cooler. The "organization" of prelife perhaps followed a gradient, in which the thermodynamic flow was from hotter to colder. The molecules had to replicate with higher copy fidelity, so perhaps the answer was to absorb heat. This absorption of heat then changed, once Darwinian selection began later.

Clearly, thermodynamic direction is also a key to cellular evolution, if a cell enclosure is a metabolic and *irreversible* process. Chemical and thermal energy can flow over a boundary, to increases order inside the cell. Cellular evolution is continuous, in that if order inside a cell is not renewed the cell 'dies', and the process ends. This is different from a *reversible* process such as information coding. A string of molecules can transfer information in non-metabolic ways; such as if a cell dies its DNA can retain transferable information.[25]

Again, if anyone disagrees with this, I ask them to reconsider facts.

Prelife began, 500 to 100 million years prior to Darwinism, using a simpler type of selection. When Darwinian selection began, at 3.7 to 3.5 million years ago, the two types of selection combined. Pre-Darwinian, *molecular* selection, cannot have gone for 500 to 100 million years, to be replaced by Darwinian, *cellular* selection, when both are different. Instead, one added onto the other, but a change is needed. It is not the mathematics. The theory must be changed, to allow for how the *molecular* selection kept going, but now *cellular* selection was added. All the subsequent paradoxes, the chromosome, species evolution, and the evolution of sex, cannot be solved without this change.

This chapter was written earlier, when my thinking was dominated by the three trees of life. This is now revised. There was no LUCA and the two trees of archaea and bacteria were for the first two billion years. More emphasis should be on how the lower life developed, as this has major effects. The next feature was the evolution of eukarya, now shown in the next chapter. For now, accept the latest trees (as given if Fig 1.3.2) as it is shown to have occurred.

---

[25]An *irreversible* process, occurs when thermal order flows across a boundary, such as a cell, but the reverse flow is not possible, without losing more order. By contrast, reversing the letter sequence of a code, such as DNA, would not change its thermal order, so it can be easily changed back. These differences in an equation cannot be combined.

This chapter is about the flowering of Darwinian selection.

It follows the early chapter on pre-Darwinian selection. The issue is that life unfolds along two pathways, of cellular and molecular change. The previous chapter described molecular evolution, but in this chapter, we must add in Darwinian, cellular selection. In this, if a scientist is trying to solve the paradoxes on a single pathway, the results will not work. To start from a single pathway to claim that it mutated into life, cannot work, or even solve how the chromosome evolved. To take this incomplete model, and project it to the evolution of sex, with a double chromosome, longer proteins, life and death, loss of horizontal transfer, will not work. It even overlooks the process of mitosis mutating into meiosis. Instead, you need the two axes for the math to add. Only once this innovation is added, do the two finally come together.

This is where mathematics is needed.

It is like in physics. You cannot say that the universe expanded only as per Hubble's Law, because there was another force, to give the shape of expansion. Similarly, you cannot say evolution began with Darwinism, at 3.7 to 3.5 billion years, and ignore how before evolution was laying down its shape, for a subsequent expansion. If you take this 'easy' route, you end with the other paradoxes, such as the evolution of the chromosome, species, sex, and the *neutral* theory of selection. It is better to correct the model from the start, and other problems can be solved.

Let me now go on, to state about the gravest paradox in the history of life, the evolution of sex.

## 2.5 The Paradox of Sex

The most famous paradox in evolution is that of sex.

If a gene from lower life, archaea or bacteria, ties to reproduce, it duplicates its genome, a 1 to 1 reproduction, with 100% of genes passed on. If a gene in higher life, eukarya tries to reproduce, for the first billion years it could, via mitosis. After a billion years, sex was added by mitosis adding on meiosis, so now two parents were needed for an offspring. It was a 2 to 1 reproduction, with 50% of genes passed. If eukarya can first reproduce without adding meiosis onto mitosis, why is this extra change needed? There are theories, explanations, books, and exceptions, but the paradox of sex is a question, why?

This book offers new explanations, not considered before. First, there is a gain along the far grey line, with the addition of *molecular* selection to *cellular* selection. Next, it was not noticed, there was a major change when sexual selection adds to the diversity of clades. Neither of these effects has been easy to work out.

If you study Fig 2.5.2, it shows a 'jump' from a bacterial chromosome to the eukaryotic chromosome, but it cannot have been that simple. Most likely, the first split from a single circular chromosome to the horizontal chromosome was when archaea ingested bacteria, at least by 2.2 billion years ago. Then between the 2.2 to 1.2 billion years, the chromosome had doubled. Sex occurred at 1.2 billion years. Other attempts at eukarya into assume that sex evolved early, so they cannot confirm how the changes occurred. For sex, we must prove why eukarya evolved, and prove why a billion years later, sex was added.

The crux to sex is the evolution of the back curve in Fig 1.2.2, because no change of the curve at the front alone can do it. Only a manipulation of both curves can achieve this. Only by a curve of the gene mutation will the front curve first drop, then rise again to the back curve. After the evolution of sex is added, the curve will then show that the front curve does not drop immediately, but continue to expand, focusing on the neutral rate of gene spread. No other manipulation of one curve can achieve this, of both the gain for sex and the neutral rate, in one theory.

Well, why did the higher cell start?

It seems that archaea needed to reproduce by enclosing bacteria. Instead of two reproductions, archaea included bacteria, so both types reproduced as one. The type, eukarya, not only reproduced as one, but it lived, gave birth, and died. Fig 1.2.2 existed from first eukarya (maybe *diskagma*). The first eukarya could reproduce, but it allowed production in a narrow line. The back curve allows a gain for sex to spread across many types, but the first eukarya did not widen its spread.

A further problem is why did genes for eukarya evolve widely at this stage. My theory is that if genes could reproduce around $6 \times 10^{-9}$ mutation, they would spread along the far gray line. This would encourage genes to hold this mutation for it to be high. Once the first mutation occurred, it would continue for all the other sequences.

Another problem is the unrooted tree of life, shown in Fig 4.3.1.

There are versions of this tree, often they are circular, but again they assume that somehow, archaea, bacteria, and eukarya evolved at the same time. Well, it is not true about lower life evolving with eukarya, but it is also not true that meiosis evolved at the same time as mitosis. Archaea tried to develop a higher organism, but this needed an input from bacteria. A newer version is shown in Fig 2.5.1. Without this tree, the affects to follow would not have been clear. Except again, this tree is not correct. It seems certain that archaea and bacteria evolved at a similar time. Evolutionist admit that the tree is faulty, and the evolution of life, show that higher life evolves later, starting by about 2.2 billion years. People admit that this is a modern tree of life, showing modern effects, but no one admits about a deeper fault at the start of the tree.

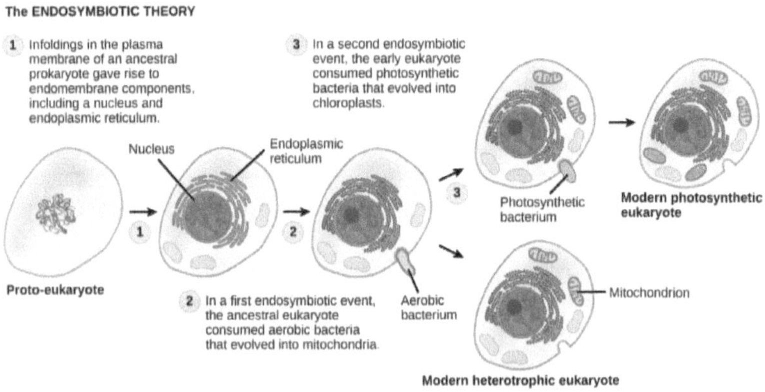


Fig 2.5.1 Another (via Wikipedia) copy of the tree. The 'Origin of Life' is shown as a blob, with no pre-Archaea or pre-Bacteria. Notice as well, there is no origin of sex. Instead, from the fist blob separate this grows until at a point the sequence splits to give first two spate lines, which then evolve on their own.

There also must be some proof why life evolves. We cannot claim it is mere fitness, because for molecular selection fitness rises, but often for cellular selection fitness falls. More of a gain is for molecules and genes to gain thermodynamic advantage. In early life, it is an advantage when genes form a circle, than when they are straight. It is an advantage, when molecules line up, on a left or right axis. It also pays molecules to form circles, of ribosomes, or elongation factors, for the base core. If molecules are forming a set of pure forms, say, of RNA or DNA, this is more of an advantage than mixtures, of random carbon chains, or non-pure RNA and DNA mixtures. Again, there were several stages, but none is firm. It might have been *grypania*, but an older organism, *diskagma,* was from about 2.2 billion years. The first eukarya spread, but nothing branched until 1.2 billion years, when sex was added. It seems that while eukarya spread, there was no diversity until sex evolved.

There is also how early cells were composed. If you look at modern genes, some are common, across all three domains. Yet where did these come from? If genes are a product from Darwinian evolution, they must have formed after this began, and then horizontally transferred across. However, there is evidence that a split between archaea and bacteria had occurred before 3.5 billion years. For instance, elongation factor proteins translate mRNA into peptide strings (proteins).[26] These can take several forms, but 1– α in humans, shares parts of the sequence across archaea and bacteria. Notice, sequence DAPGHRDF-KNMITG-SQAD-A-L-V is highly conserved for the human and archaea domain. It is more likely that this was pre-Darwinian, and then split once Darwinism occurred.

| | |
|---|---|
| Human | DAPGHRDFIKNMITGTSQADCAVLIV |
| Archaea | DAPGHRDFVKNMITGASQADAAILVV |
| Bacteria | DCPGHADYVKNMITGAAQMDGAILVV |

Table 2.5.1

I suggest that higher life, eukarya, evolved by 2.2 billion years ago, from lower forms. It followed its evolution in a narrow line until the evolution of sex, at 1.2 billion years. It seems that the organisms died, for whatever reason, from 2.2 billion years ago, but that sex evolved at 1.2 billion years. It is not good, but you see a straight line from first eukarya, at 2.2 billion years ago, to the evolution of sex, at 1.2 billion years ago. This sequence needs clarifying, but this is how I interpret it.

I was therefore surprised by Nick Lane, and others, who then put the evolution of sex down to the inclusion of mitochondrion. I think this is an impressive argument, which was not included. Despite the impression of it though, it firstly, cannot explain the results of sex, the way I have done. Then Nick Lane and others mixed up the actual timing. To me, eukarya, without sex, evolved for about a billion years before sex was added. It seems obvious to me that mitosis existed a long time before the addition of sex. Over a million years, perhaps many features of mitosis were shifted

[26] The DNA code is translated first into (messenger) mRNA, then into proteins. Ribosomal rRNA and transfer tRNA uses its own code, which is not copied from DNA.

from the mitochondrion to the nucleus, an important effect that might itself have led to sex when no more shifting could occur. After the billion years, however, sex was finally added to the nucleus. Excuse me Nick Lane, and others, who have worked so hard for an explanation of sex. However, my theory of the two axes of spread can explain the result. It can also explain the neutral rate, missing in the other explanations.

So now, consider the evolution of sex.

There was a mix up, even in my own thinking, between the evolution of the two pathways of gene spread in Fig 1.2.2, and the genetic pathways. Somewhere, the gain for sex produced the diversity below, but this might be a Darwinian adaptation. If you consider a straight line to the branching for sex, the proliferation might be mixed with the original.

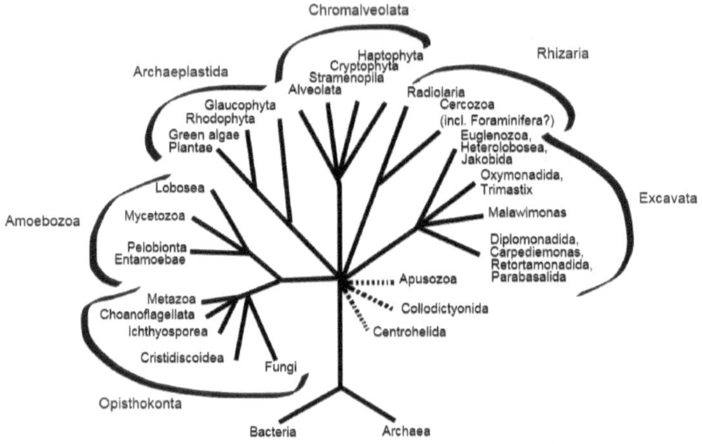


Fig 2.5.2 A copy (from Wikipedia) of growth of diversity over the domains. This diagram can be revised, but notice the diversity comes roughly, after a billion years prior, of slow adjustment in the first evolution of Eukarya.

Rather, a single curve along one pathway cannot show the gain for sex. There must be two gains, in which the back curve shows the mutation rate. I have been questioned why, because by $x_i$ the gene is more conserved with a higher mutation rate. However, I request that experts check gains across life. Only with both curves shown, can we see why there is this gain where the results show the gains for sex.

The other issue is that even after sex, genes have a horizontal transfer, which has no explanation in the standard theory. From manipulations of the front curve alone, no matter how clever, there is no mechanism of why the gene goes from the front to the back, by any type of horizontal transfer. However, with my theory it will. Check Fig 2.2.2 again. If there is a gene which evolved after sex, but it needs to migrate to the back curve, it can still 'horizontally transfer' by leaving the normal after sex route, and directly transfer to the back curve by moving outside of normal sex (it can appear as a dotted line). I have not checked this directly, but it occurs, for these types of genes.

The final problem is how did genes change their manner of distribution, once eukarya developed the addition for sex. There is no proof about it in any theories, but it would be a test of sex. My theory, it is only that, is that genes for sex shifted 'westwards' as they were driving. Without extra data it is hard to determine what happened. However, as the back-curve falls, the gene moves towards the Darwinian spread. Yet this is arrested by genes keeping the driver for sex longer. Roughly, the genes keep a sex driver by holding the spread to about 0.71 of gene spread along the far line. I am not certain how, but a change affecting genes keeps to back line from rising too high. This could account for the gene spread of around 6 x $10^{-9}$ for molecular theory.

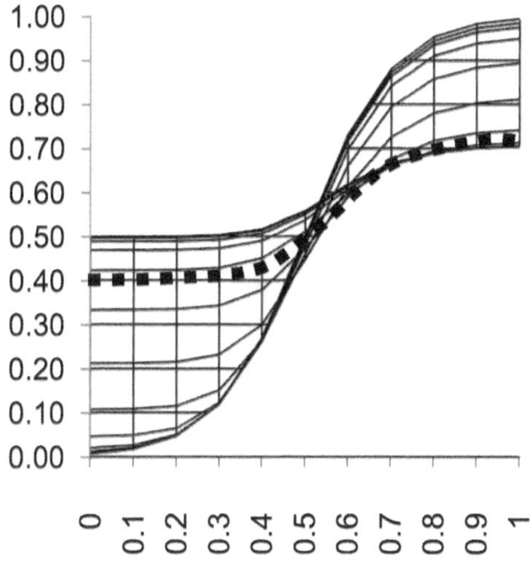


Fig 2.5.3. Earlier curve modified to show the *neutral* point of molecular selection. A curve of pure Darwinism goes from zero, to peak at 100% on the front curve. This limits the gain by sex. However, if a gene can stretch out its *neutral* rate, so it appears halfway along, it can also gain by sex, so that it splits each way.

Whatever the pre-sexual problems, archaea and bacteria overcame it. When these combined, the new chromosome had a horizontal form, rather than circular (though one bacterium for mitochondria stayed circular). The combination might have been each gene vertical, but by sex, two genes formed the vertical, with one gene for sexual determination.

Once sex evolved, at 1.2 billion years, it dominated all the new forms to evolve beyond it. To evolve from archaea and bacteria into a eukaryote was complex. The genome had to increase by 1,000 times, and change from a circular chromosome to a double chromosome in eukarya. There was also change so that copying in lower life ceased, to become a parent-child set up. If you ask the experts, they have no explanation of the synonymous DNA rate, and none that includes genes able to spread for sex. Yet if you study the rates, it seems to be the case. It seems that in the first eukarya, conserved genes such as ubiquitin, H4 and others could spread, even if not widely. Sex not only was a change, but it brought an effect such that genes needed for sex could increase mutation by use of synonymous DNA. This can repeat parts of DNA that do not express proteins. Table 1.4.1 shows this pattern. This replicated, but it had little diversity, until sex evolved at 1.2 billion years.

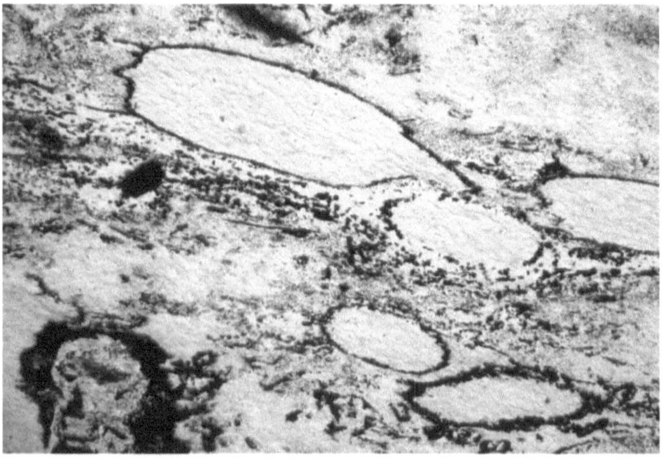

Fig 2.5.4 Photo of *diskagma,* that evolved some 2.2 billion years ago.

Notice too, all the other theories of sex use math, but they assume that the gene spreads along the Darwinian, *cellular* selection. None allows that the gene spreads along two axes, molecular and cellular, and the molecular axis is for genes to develop new clades. The previous theories try to prove how a fall of cellular selection from 100% to 50% was a gain, but it does not work. Instead, cellular selection was from 0% to 100% for any species, with or without sex. Molecular gain was 100% for the first histones, but it fell from there, to 71% for full gene spread, as a *neutral* rate. If gene selection falls below 71%, moreover, it will revert mostly to Darwinian spread only, show in earlier diagrams.

Without molecular selection, the theory of evolution is incomplete at any phase. Fig 1.5.1 is typical, but not accurate. The animal species, the most numerous, might not have formed until half a billion years after the first branching. Plants also imported from bacteria. If sex formed at 1.2 billion years, various branches added after that. If the chromosome divided in two to allow wide spread, this act alone does not explain the increase in diversity. Instead, in time, the two chromosomes allowed for a wide increase in diversity as the genes spread.

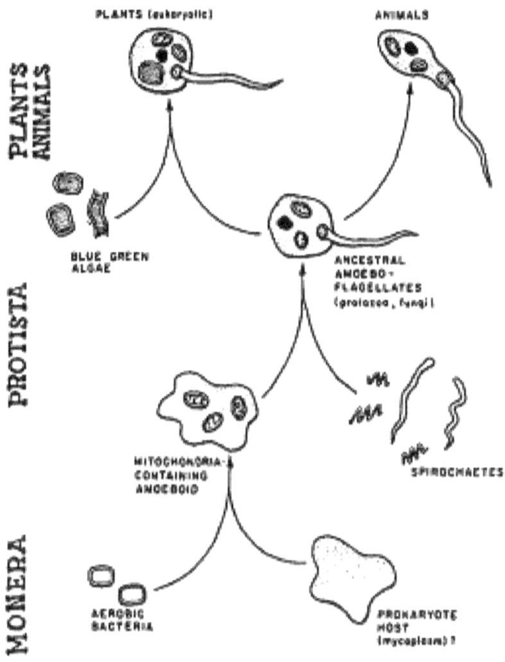


Fig 2.5.5 This is from Wikipedia, for the evolution of eukarya. It is from Margulis, but the spirochetes are not considered correct, and the evolution of mitosis into meiosis is not shown. The double chromosome evolved with Protista.

In Fig 1.5.2, there is a huge increase in diversity at 1.2 billion years. About six new clades and several minor ones arose. However, there was a drop of cellular gene spread, from 1:1 down to 2:1. How could a sudden drop in cellular selection, to 50%, at the same time cause an increase in diversity. There is no mathematics from *cellular* change to explain this. A rate of mutation above $10^{-13}$ lets the gene spread widely. However, then sex came, the gene could spread way down to 0.5 to 0.71 % of the fitness (see Table 1.6.1).

The double chromosome gives a two-order gain of parity (like from single-stranded RNA to double-stranded DNA). The new chromosome evolved for sex, but archaea and bacteria exchanged by conjugation. Frequency appears after gametes specialized as male and female, but not each case of isogamy evolved into anisogamy. Moreover, a 50% loss is only the start. In meiosis the gene duplicates twice, which leaves 3 out of 4 genes lost, a 75% loss, higher than the 50% loss. [27] It does not mean that sex, or that *isogamy* only evolved into *anisogamy* (see footnote). The evolution of sex cannot occur unless mitosis first mutates to meiosis, even if the male-female is not always followed on.

It is correct there is no other explanation based on cellular selection, but even with molecular selection, change of mutation rate is strange, especially the synonymous DNA variation. However, consider the other explanations of sex. If you study Fig 1.5.1, we think of diversity via a "series". That is, one branch leads to another, and then more after that. Once sex evolved, though, diversity appeared "horizontal". For a billion years, there was no diversity in higher life. If cellular reproduction is in a "series", then diversity is a "series". With sex, it seems as if the gain was in parallel, so the increase in diversity is horizontal.

Instead, there are two types of selection. If you combine molecular and cellular selection, a combination does produce a gain. You can take these two types and test it. Notice how with the two gains an increase for sex is possible. On the new curve, regardless how the front curve responds, the back curve will increase if the mutation rate stays high. If a gene for eukarya starts with a $10^{-13}$ (or slower) mutation, it will spread across higher life. Once sex has evolved, a gene increases mutation from $6 \times 10^{-9}$, the gene will spread in higher life. If the gene increases mutation greater than $6 \times 10^{-9}$, it will lose direct gain. Mutations more than $6 \times 10^{-9}$, the new gene will follow standard gain at the front.

Now consider how the two distinct sexes arose. Again, I have no proof, but suppose the chromosomes doubled with sex, but the autosome (of one gene) chromosome stayed separate. This continued until the allosome (for sex) divided into species, which would lead to male and female. This first occurred when archaeon ingested a bacterium. The two duplicated before the cell divided, and was about 2.2 billion years ago. When meiosis was added, another split occurred at 1.2 billion years ago. The problem is that the chromosome could not jump from circular to a double chromosome, without intermediate steps. Once the first ingestion occurred, separate chromosomes were needed, perhaps one for archaea and one for bacteria,

[27] Strictly, 'gamy' refers to gamete size. Here I use 'gamy' to denote if the sequences being exchanged were identical or differentiated, in the parent chromosomes.

but with a way to replicate. Splitting of the chromosome allowed this, but a billion years later, extra complication needed the double chromosome, by allowing sex.

Again, there is no theory offered of how it developed. With *diskagma*, or similar, parents had offspring, and the parents died in time. The existing theory is that the doubling of the chromosome was also early. However, my suggestion is that it might be later, once meiosis added on. If this, single chromosomes might have gone for a billion years, before meiosis was added on, with sex now showing a double chromosome.

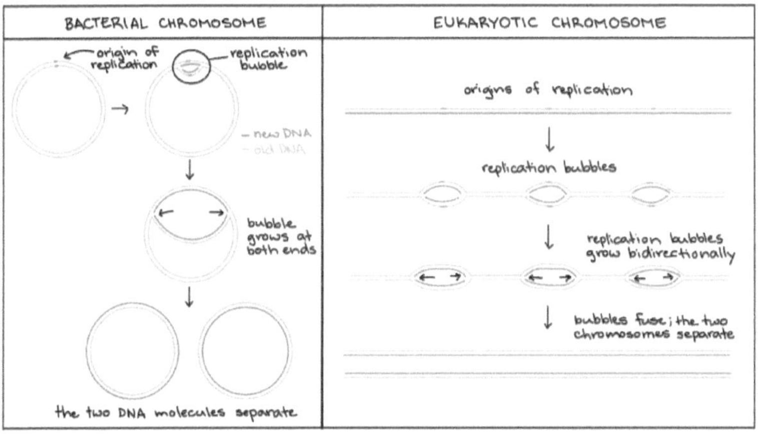


Fig 2.5.6 In the lead to sex, the split from circular to the groupings was from about a billion years. Sex with meiosis was in the final stages, at 1.2 billion years.

Now, in the standard theory there is a dispute if meiosis added onto mitosis after a long period. Again, no one seems sure, but an assumption is that it took a billion years for the mitosis and meiosis grouping. We cannot be certain when the second chromosome joined, but only that sex evolved about 1.2 billion years ago. It is hard to see how mitosis could be a long time, but meiosis results from a different cause, such as bacterial division. More likely, bacterial genes diversify by horizontal exchange, but this was not possible once pre-eukarya evolved.

From a first four-fold reproduction, eukaryotes could start to branch, or those that emulated a four-fold reproduction survived. Other attributes of sex, where isogamy evolved into anisogamy, DNA repair, a two-fold cost, or loss of sex in species came later. Even in anisogamy chromosomes retain 50% homology. It would be interesting to test genes such as hemoglobin, how much of the core gene was conserved across many species. Genes were also suffering a 100% loss from the start, so the surprise of sex is a 50% loss as genes passed on.

Rather, consider sex, as frequency gain for genes, until the neutral rate. Note, the '1' in the bottom curve keeps the rate high at the back, but later this reverts to a standard front curve, as conservation falls further. The *neutral* rate is an effect, where the genes can gain along both axes. Now it might seem from Fig 1.5.6 that the gene will lose frequency going to the back curve. However, assume that a gene tries to gain frequency, and it will pay the gene to gain along the two axes first. There are other theories, including a two-fold cost, resistance to parasites, deleterious genes, genetic variation, novel genotypes, and the speed of evolution. If experts, even if they have written about sex, if they have tried the other equations, still confess that they do not know why sex evolved.

Darwinian, cellular selection works by, call it, a "front black curve". There is also molecular selection, and it unfolds to, call it, the "far grey curve." The problem of sex is to get genes from the "front black curve" to the "far grey curve" without a loss of fitness. Notice however, there is an inevitable loss, to about 71%, on the right side of the model. You cannot escape this loss of fitness. Of course, this does not say everything about sex. You might wish to check the other theories of sex. You might decide to investigate species that had sex, and since lost it. Some minor species might have slipped through without sex. Other species, insects, flowers, and other types have weird varieties of sex. If nature is a tinkerer, this is how it works. The math also works. Genes maximize spread along a *neutral* rate, but once conservation falls below this, the figure will revert to the front curve.

Still, you must check why genes reverted to non-sexual reproduction, and theories why sex is averted in a few species. One asexual organism, Bdelloidea, had a quadrupled chromosome. A further problem is that the difficulty causing sex must have applied from lower life. Early genes of RNA were 50% conserved for billions of years, but as the "sub" lengths of a chromosome. If life began by genes joining, sub-length to sub-length, conserved RNA laid the short lengths. If a need for genes to spread widely is a key to sex, it extends a problem that must have existed since then. Again, early genes kept about 50% homology, so that once these evolved, they held close to or above 50%, or $\varepsilon_k \geq 1$.

Of course, this could be wrong. There is no absolute proof that the first eukarya was *diskagma*, nor the diversity was the addition of sex, or that the chromosome doubled for sex. Rather, an effect at the start of sex, leads to a change in diversity. If it were *diskagma*, from 2.2 to 1.2 billion years, there was little change over the period. Some effect caused the sudden diversity in eukarya, once sex evolved. This too was likely the addition of meiosis. The evolution of mitosis was already complex, so an addition of meiosis on top was an extra factor.

Now, let me add on the CRISPR/Cas9 theory of gene modification. First, this is not a rival theory, but is a feature that adds onto the existing model. First, there has been a long discussion of how the CRISPR model of gene spread applied. It looked useful, but with no model of how to apply it, until the CRISPR could be added to the Cas9 sequence. This has been developed, and now there is a range of CRISPR/Cas9 models or similar that can be used. This is now an extensive new method of how this system can be applied.

This section on sex has been rewritten, to include several effects not considered until now. The main effect was the increase of mutation rate along the back curve. If you try to prove the gain of frequency by the front curve, it will not work, especially from a 50% loss of gain by genes passed on. Only the two curves, with gain along the back can solve this problem. The gain for sex must show an increase in mutation rate to be sustainable, and explain the *neutral* rate.

The next point is the close relationship between the evolution of sex and the time that diversity occurred. The evolution of eukarya, perhaps it was *diskagma*, began at 2.2 billion years, but with little diversity. Sex, and with suddenly with huge diversity in clades, began at 1.2 billion years. It seems incredible experts could write books, even on sex, but not mention this huge change. The change was also roughly from "serial" into a "horizontal" diversity at that point. Again, incredible, people write about sex, but not even mention this. The final change was from archaea and bacteria into a new type, eukarya. It was not just a new type, but eukarya lived, grew, reproduced and then died. Eukarya was about 1,000 times greater than the previous forms. Eukarya began at 2.2 billion years ago with the reproduction of *diskagma*, or similar, but sex began at 1.2 billion years, and with all the added complications.

It might also seem strange that in all of evolution there is no gain for mutation rate, but think of the changes. There was gain for first selection, before cells. There was gain as RNA changed to DNA. There was gain as lower life changes to higher life as eukarya. Finally, there was a gain as no-sexual reproduction became sexual. Why would all these occur at specific points, if there were no gain?

If a gene had 100% gain on the front peak, it could not then spread to the back grey line, unless it lost frequency. It must fall to the *neutral* point. Now, the first proteins to spread in higher organisms, had ubiquitin from bacteria, and histones are from archaea. These would evolve pre-sexual. Rather, the problem is for "sexual" organisms to evolve. If these were successful, and spread for 100% on the first peak, they would be trapped. To reach the back curve it must transit the right side, to cross "dip" between the front and back.

At a point, the beginnings of sex evolved, by two organisms making this change as the basis of mixing. This was after horizontal exchange to increase gene spread in lower life mostly ceased. The differentiation into male and female came later, as not all organisms made this final step. However, it is impossible to see how this step could be made, unless genes competed along two axes.

If experts accept that genes gain in frequency across life, and the two types of gain combine, the math of the gain is obvious. If *cellular* gain is logistic, and *molecular* gain is logistic, or similar, the combined curves reproduce a "dip" on the right side. From this "dip", obvious, but not considered before, there is a major complication in evolution of the gene, not just for sex, but also for all of life.[28] Molecular gain is selective, but it will appear in the equations along a new axis.[29]

If the base model is understood, the evolution of sex can be solved, not just the 50% loss, but all peculiarities across life.

---

[28] For pre-life, there is an Eigen value model. For the chromosome, there is the hyper-cycle. For sex, there are many models. Modern models are also co-variant, with extra factors.

[29] Some people do not understand 'imaginary' gain, but it is the gain of conservation occurs "before" a modification by mutation. From first life, it is easy for genes to copy accurately, "before" cellular modification, to set the genes to fit in with overall theory. Note, every gene in eukarya must evolve against ubiquitin, histones and others.

## 2.6 The Neutral Molecular Theory

How does modern, molecular theory work?

Well, there is a calculation, of the nonsynonymous to synonymous ratio in a gene, or the $K_a/K_s$ ratio. Change of the $K_a$ ratio influences selection, but the $K_s$ change will not. In the standard model the advent of sex does not alter the $K_a/K_s$ ratio. Rather, this is used to calculate the neutral, near neutral, or selective pressures on evolution.

By contrast, my theory considers that the $K_a/K_s$ ratio will change with sex. It will shift the curve "westwards". Unfortunately, without MATLAB, I cannot show the full effect, but with sex the neutral rate does not fall so rapidly. Instead, the advent of sex pays the gene to maintain the neutral rate extensively. Instead of simply falling off, sex allows the gene to keep the rate high, so a mechanism in the gene can simulate a high $K_s$. We can notice this in the results. Prior to sex, the $K_s$ ratio would fall, but with sex, the $K_s$ ratio can extend, such that modern genes (from sex onwards) can exists as though the ratio continues.

Well, those are two theories, but which one is correct? It cannot be that hard to test both theories.

Well, 4.1 to 3.8 billion years ago, life started from a different theory. There was no Darwinian selection, and genes or pre-genes that copied the most accurately spread the widest. The early competition did not cease once cellular selection began 3.7 to 3.5 billion years ago. Molecular selection would continue, as a background to Darwinism, to modern times. If this were true, how would this idea, which is different from the standard, compare to my new molecular theory?

The problem, let us call it, is the *neutral* theory.

Now, I am not sure how it began. If the only means of selection is by Darwinism, most examples of fitness will either go down, or be *neutral* or non-selective.[30] This was called 'non-Darwinian' theory, though it was changed to the *neutral* theory, after problems with the Creationists. My opinion, is that experts were investigating the punctuated rates, but when they investigated this against Darwinian selection, it mostly went down. This is how the modern term arrived.

Anyway, we ended up with the *neutral* theory.

One interpretation is that the neutral theory is an existing problem that was incorrectly interpreted. There is a fixed neutral rate, which has arisen at least by the evolution of sex. If this is true, then sex will favor genes that mutate close on or above the neutral rate. The calculations will work. My theory is that the more conserved a gene since it evolved, the wider it has spread across the populations of life. For example, in eukarya, the most conserved gene is ubiquitin, followed by H4, then H3, and down a scale,

---

[30] It solves a problem in the next part, on how Darwinian selection often seems to fall.

from $10^{-14}$ down. These are the widest distributed. At the other end are virus genes, from $10^{-3}$ up. These are the shortest distributed, and can even be down to just one species. We can trace this across life.

This new theory is tested with *molecular* selection, using the formula that has applied since such selection began.

$$X_k = \varepsilon_k/(1 + \varepsilon_k) \qquad (2)$$

The contrary view is that of the experts. They claim that there is little *molecular* selection. The only way instead to test a gene is using *cellular* selection. Using this, we must test gene spread with the formula;

$$\Delta x_i = x_i(w_i/\bar{w} - 1) \qquad (1)$$

In this, ubiquitin, H4, then H3, and so on, are the most conserved, but these are not considered. Instead, the results focus on an effect further down the scale. As we get closer to the *neutral* rate, it pays genes to increase this rate by increasing the uncoded sequences. This gives a rate of 6 x $10^{-9}$ for typical genes. This leads to a theory of *neutral* selection, in which terms that might apply to genes, such as 'non-Darwinian' are ignored. Experts can calculate from it a *neutral* rate. Say, other effects higher up the scale are then ignored.

Now, apart from the math, my theory includes a prediction, that the more conserved a gene, the wider it has spread across life. The expert theory also has a prediction, that there is no correlation between gene spread and gene conservation. Instead, these results are further down a scale, and genes distribution correlates only to the neutral rate. Both these theories should be easy to test. They cannot both be right, so which one of the theories is correct?

Well, we must be careful. The expert theory did predict an important result of gene spread around the *neutral* range. To reach this, it studied the synonymous DNA (which does not affect fitness), which tends to a rate around 6 x $10^{-9}$ mutations. My theory looks at the nonsynonymous DNA (which alters fitness), which varies across the base rates. I have suspected that the synonymous DNA rate is an adjustment for sex. This is such that a high synonymous rate adjusts so that the DNA sits on the curve (see Table 2.6.1) by adjusting for sex.

Another problem is that genes spreading across life in proportion to conservation seems odd, in that the longer a gene existed, the more conserved it is. One reason is length. Say, the most conserved gene in eukarya, ubiquitin, is also the shortest, and the next, H4, is short as well. Another is how longer genes have more uncoded sequences. As well, conserved genes must have purifying selection, such that they abort if violating high conservation, even in the womb.

The next issue, the H4 histone is spread across higher life, it should be $X_k = 1$, but using the formula, $X_k = 0.99$. Other histones, H3, or H2A and H2B, are less conserved, but they still should give $X_k = 1$. Others are difficult too. The Hox gene should be $X_k = 1$ in animal and fungi, but drops to $X_k = 0$ in other species. The Hox has nearly the same sequence, but it is arranged differently in different phyla. Another problem is for genes such as cytochrome c. It should be $X_k = 1$, in eukarya, but it is further down the chain, and varies across species. Often the mutation rate is not clear, or the gene spread is not available. $X_k = \varepsilon_k/(1 + \varepsilon_k)$ is a mathematical interpretation of a problem.

Still, taking Equations (1) and (2), they give the same output, expect Equation (1) is for single species, and it includes fitness. By contrast, Equation (2) measures genes conservation across all life, but it does not included fitness. To be fair, the experts did not have Equation (2). It did not exist, when the *neutral* theory was formed.

There was a controversy over the *neutral* results, or if terms, such as 'non-Darwinian' evolution were used. Once Creationists learned of 'non-Darwinian' effects, the terms were modified. Now, *neutral* became the official terms, but even then, there was controversy. There is still a non-Darwinian result, and yes, even now the *neutral* rate is not agreed. Apart from that, some eukarya genes, such as ubiquitin, H4, H3 and Hox genes are highly conserved. Again, the problem is that an incomplete theory had taken a top spot. In the end, the standard evolutionists had to accept the *neutral* theory. It had always seemed wrong, but here the experts in mathematics, held the sway.

After thinking about this, a long time, the new theory has a model of *neutral* selection, which also reflects facts. The next Fig 2.6.1 shows the model. The issue is that without a model of *neutral* selection, experts became focused on a different problem. In the "old" theory of evolution, for every gain of Darwinian selection, fitness must rise, but that cannot work over a huge scale. For human evolution, say, we cannot claim that fitness only rises; this would lead to a catastrophe. We cannot claim that humans, which evolved later, are more fit than worms, which evolved first. Rather, experts had to investigate, how Darwinian selection worked. They found that for every Darwinian success, there were more failures, where selection went down. They also found that changes were *neutral.* A random change of genes often had no effect on fitness. It was non-registered, did not affect fitness, or features such as skin patterns might change, but also with no change of fitness.

Rather, molecular selection does rise as life evolves, but along a new axis. This rise has complied since first life. There is also a solution to the *neutral* selection, which can solve the previous paradoxes, such as the evolution of sex. Again, you need to base this on a different axis of gain, by an imaginary number for molecular distribution.

To prove the second curve, assume that the average mutation is about $\overline{\mu} = 6 \times 10^{-9}$. (It is near $10^{-7}$ in lower and $6 \times 10^{-9}$ in eukarya.) Use it for mutations, $\mu_k = 10^{-14}$, $10^{-12}$, $6 \times 10^{-9}$, $10^{-6}$, $10^{-3}$, in modern species. Then derive these values for $\varepsilon_k$, and $X_k$, in the following table. Again, it looks mathematical. Actual rates also vary, depending if you included only higher genes, or both lower and higher genes. Mutating virus genes also drag rates down to lower values, if they are included.

| **Gene Case** | $\mu_k$ | $\varepsilon_k$ | $X_k$ | **Comment** |
|---|---|---|---|---|
| Highly Conserved | $10^{-14}$ | 68.3 | 0.99 | Near 1 |
| Conserved | $10^{-12}$ | 9.2 | 0.90 | Highly conserved |
| Neutral | $6 \times 10^{-9}$ | 1.0 | 0.5 | Neutral Rate |
| Fast Mutating | $10^{-6}$ | 0.02 | 0.02 | Fast Average |
| Very Fast | $10^{-3}$ | 0.00 | 0.00 | Near 0 |

$$\varepsilon_k = e^{(\log(\mu k) - \log(\overline{\mu}))}$$

Table 2.6.1

The *neutral* rate tends to be longer, as there is a benefit to higher genes to settle near this rate. Lower genes do not have a *neutral* rate. Some short RNA and DNA sequences can be near infinitely conserved since first life. The histone H4 is under $10^{-14}$, whereas a few virus genes are near $10^{-3}$. You can graph these plots, and the longer that a gene existed, and then the more conserved it is, across life.

Now, to get a plot, I had to mix ancient and modern values of genes. I do not have reliable plots of gene values in prelife, apart from noting, say, that the 23s Ribosome is near infinitely conserved (as molecules), as it spread across lower life. However, the mixing of modern and ancient values makes my case stronger. Remember, experts on evolution, state that there cannot be a theory of *molecular* selection, however Table 1.6.1 shows the theory is there. The more conserved the gene, the wider it has spread across life. If anything, I can prove how the existing theory does not fit with how evolution works in general, and the expert view is not correct. Rather, a different issue is stopping how *molecular* selection interacts with life.

The next curve illustrates the approximate shape. This is the curve in Fig 2.2.3, but redrawn. The front curve shows the Darwinian selection, ending on the front peak. Then the next, dotted curve, illustrates the modification for the *neutral* theory, halfway along the curve to the back. Notice, if a gene had evolved so it could obtain the *neutral* point, the *neutral* gain would allow it to diverge across many individuals. Genes that are highly conserved would spread across to the back line, and this continues down to the *neutral* rate. For genes conserving sequence below the *neutral* rate, the curve reverts to the front diagram only. This result, further confirms the theory of the two-axis spread.

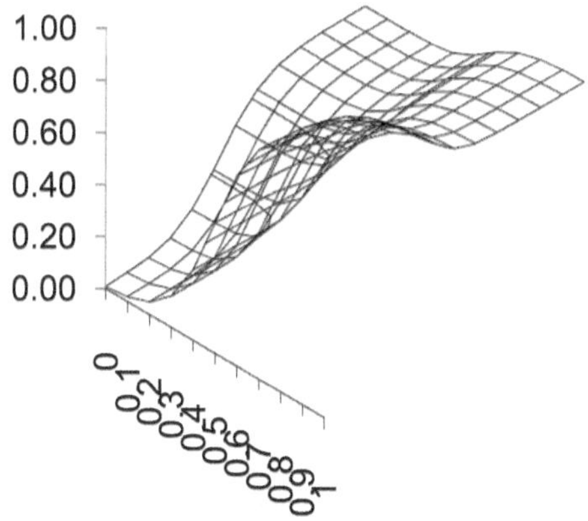


Fig 2.6.1. Earlier curve modified to show the *neutral* point of molecular selection. A curve of pure Darwinism goes from zero, to peak at 100% on the front curve. This limits the gain by sex. However, if a gene can stretch out its *neutral* rate, so it appears halfway along, it can also gain by sex, so that it splits each way.

Notice as well, although the new theory is different from the existing one, many facts of the existing theory stay the same. Genes still keep a *neutral* rate, except that now it occurs along the back curve, but this has other advantages. While the *neutral* rate is in the correct place, it also rises where genes gain frequency. For example, ubiquitin, histones, and the hox genes all gain greater spread. The genes also gain greater spread across life, back to prelife. Genes can also account for paradoxes such as the evolution of sex.

If we accept that genes can sort across life, by the rates at which they mutate, we need to relook to how this began. Pre-cellular replication was for conservation of sequence. Once cellular evolution began, RNA began to combine the mutable sequences with the previous sequences. RNA is single-stranded, and is less stable than DNA, but it can hairpin fold, so pin legs form a double-strand. If the first genes were RNA, the mutation rate of RNA is often taken as $10^{-5}$. However, this high RNA rate is based on viral genes, but there is little proof that these were the first life. Instead,

RNA genes began with hairpin folding, where conserved RNA exists for millions of years, but as smaller "semi-genes" from within the RNA framework.[31] (Note, I went to trouble assembling the next few tables, but no one has commented, so for now, assume that they are correct.)

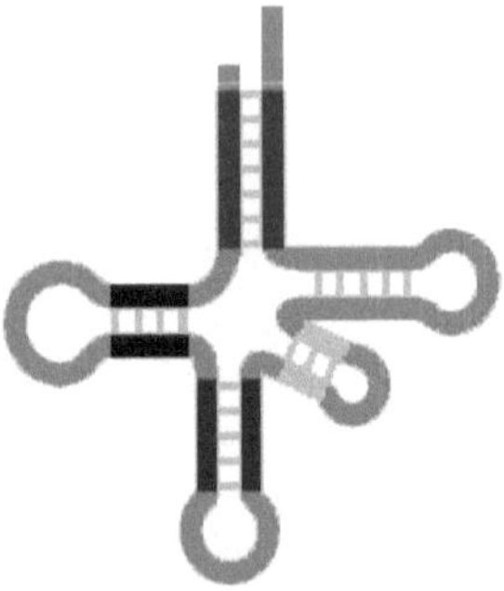

Fig 2.6.2. RNA is single-stranded, but short lengths can hairpin fold in a double-strand. The double-strand in tRNA was a likely precursor to DNA.

Table 2.6.2 shows tRNA for *Saccaromyces cerevisiae* in eukarya, *Haloarcula marismortui* in archaea, and how *Thermus thermophilus* in bacteria, are distributed across the three domains of life.

| | | | | | | | |
|---|---|---|---|---|---|---|---|
| GCGG | ACU | UAGCUCAG | UUGGG | AGAGC | GCCA | GACUGAA | GAUC |
| GCCG | CCU | UAGCUCAGACUGGG | | AGAGC | ACUC | GACUGAA | GAUC |
| GCCG | AGG | UAGCUCAG | UUGGU | AGAGC | AUGC | GACUGAA | AAUC |

| | | | | | |
|---|---|---|---|---|---|
| UGGAG | GUCC | UGU | GUUC | GAUCCACAGAGUUC | GCACCA |
| GAGCU | GUCC | CCG | GUUC | AAAUCCGGGAGGCG | GCACCA |
| GCAGU | GUCG | GCG | GUUC | GAUUCCGCCCCUCG | GCACCA |

Table 2.6.2

Notice that despite RNA's reputed high mutation rate, 50% of the above tRNA sequences have stayed 100% conserved across all the domains of life, from three to two billion years in the past.

The next part is that not only RNA doubled along the hairpin fold, but also later RNA doubled into DNA, which has a double strand. Notice, even

[31] RNA might have been the first genes, but not that RNA viruses were first life. Viruses are parasites off existing life. Still, viruses might have evolved early for other reasons.

after DNA evolved, this doubled to form the third, highest domain, for eukarya of higher life. Consider an enzyme sequence *triosephosphate isomerease* for processing glucose. Its sequence varies for a bacterium, a chicken, and a fungus. The DNA gene contains high conservation for crucial sequences (sequences underlined) but variation elsewhere.

```
MRKNIVA  GNWKMNKTLQEGIALAKELNEA  LANEKPNCD   VIICTPFIHLASVTP
MAPRKFFVGGNWKMNGDKKSLGELIHTLNGAK LSADTE      VVCGAPSIYLDFARQ
MVTTTKALFPPLPKTLLIISLKMYFPPDRTLSYLRDLLSPANKIVLPQNRSRLLLALIP

LVDAAKIGVGAENCADKESGAYTGEVSAAMVASTGAKYVILGHSERRAYYGETVEILKD
KLDA KIGVAAQNCYKVPKGAFTGEISPAMIKDIGAAWVILGHSERRHVFGESDELIGQ
DFLTIYPCAQIIKDWAAGAGAYTGEVSPASLRSLGVRLVELGHAERRALFGETDDQVAR

KVKLALANGLTPIFCIGEVLEEREANKQNEVVAAQLASVFD
KVAHALAEGLGVIACIGEKLDEREAGITEKVVFEQTKAIAD
KAAAAVDQGLIPLVCIGEVTAPGAIASEAVGLAFQLPSDVDPEQPPFLLGAQDCFWEAV

LSAEDFSKIVL               AYEPVWAIGTGKTASPAQAQEIHAFIRSAVAEKYGKE
NVK DWSKVVL               AYEPVWAIGTGKTATPQQAQEVHEKLRGWLKSHVSDA
RECAGQMRAVLDAIPSAAPVIFAYEPVWAIGKAKPAGVDHVAAVVEGIR AVIGKREGE

IADNTSILYGGSCKPSNAKELFANPDVDG GLIGGAA     LKVADFKGIIDAFN
VAQSTRIIYGGSVTGGNCKELASQHDVDG FLVGGAS     LK PEFVDIINAKH
V    RVLYGGSAGPGLWGAGGLGKAVDGMFLGRFAHEIEGVQKVVQEVEETLSEQ
```

Table 2.6.3

Once we are in higher life, some genes are conserved along the entire length. The most conserved is the H4 gene (apart from ubiquitin) which varies two letters (98% conserved) between a human and a garden pea.

```
Human: MSGRGKGGKGLGKGGAKRHRKVLRDNIQGITKPAIRRLARRGGVKRIS
Pea:   MSGRGKGGKGLGKGGAKRHRKVLRDNIQGITKPAIRRLARRGGVKRIS

GLIYEETRGVLKVFLENVIRDAATYTEHAKRKTVTAMDVVYALKRQGRTLYGFGG
GLIYEETRGVLKIFLENVIRDAVTYTEHARRKTVTAMDVVYALKRQGRTLYGFGG
```

Table 2.6.4

At first, it bothered me, why H4 could be 98% conserved, whereas RNA genes could be 50% conserved in lower domains. The answer is in how life evolved. Early genes were 50% conserved along hairpin folds, such as archaea and bacteria. Early DNA genes must have repeated limitations from RNA. Only once genes evolved into higher life, after that were longer proteins possible, for one domain.

The other point is that while the intent is how genes grew in length for conservation, it does not mean that mutations in another part of DNA did not bring it about. It was a fact, RNA was replaced by DNA, or genes doubled in higher life. Still, you cannot claim that mutation alone caused genes length to gain, so mutation was the cause of change. If a mutation caused the genome to double, the mutation was a loss of frequency, but the genome doubling was a gain of frequency.

Consider a penguin and a rose, competing by natural selection. In standard theory, penguins compete with penguins, but roses compete with roses. In the new theory, conserved genes in penguins compete with conserved genes in roses, diverged over millions of years, or distance. The eukarya species had the common genes, ubiquitin from bacteria and histones from archaea. Anything can change, but a H4 histone from eukarya, possible a *protozoa*, is more conserved than a H4 histone from a penguin, so the competition is there.

If we broadened the range, conserved genes from a protozoa, will be slightly more conserved than genes from a slime mold. If we looked at animal genes, pre-Cambrian fossils would be more conserved than post-Cambrian. Scientists accept that the more conserved the original gene, the further back it evolved, but no one can connect this as a competition. Especially, competition would require a model of molecular selection, to interface with the existing model of cellular selection.

The other problem is that if molecular selection began in prelife, and genes gained stability, it would not pay a gene to lose stability, switching to mutation. For instance, there are 50% conserved RNA genes from 3.5 billion years old, but early mutation was within a gene, for mutable sequences to combine with conserved ones. That conserved parts have spread across three domains cannot alter a perception that genes must mutate, before there is life.

Even after life began, there is no point where a gene spreading by conservation, must give up all the gains. It could not be after molecules were ingested in cells, because extra stability was needed for groupings. It could not have been after genes began encoding proteins, because early coding was not exact. More than a billion years after molecules became ingested inside cells, a further ingestion occurred: cells from archaea and bacteria formed a super-cell, *eukarya*. This led to the evolution of sex, where life branched into plants, fungi, and animals. Evolution of eukarya gave hundreds-fold gain in conservation. Yet selection for conservation did not cease, because eukarya genes that conserved sequence highest, have distributed widest across that domain.

Later we see this is why rogue fragments impair fitness, to conserve sequence. It is not that rogue or parasitic genes and DNA do not prefer a stable existence. Over the history of life, every gene had to alter until it found a sequence in its organism, where a gene can spread. Over billions of years, the prime spots in the genome were filled with stable sequences. All DNA tries to attain a stable sequence, and force other DNA to alter. Just that modern DNA will enter saturated genomes, and it must conserve sequence against earlier genes. As long as DNA is not harmful, it will not be selected out. If it alters too fast, DNA will not translate in ways to express proteins, but junk DNA,[32] like computer code that does no harm, might be harder to select against.[33]

Having outlined how life might have begun, with molecular selection first, which gave way to cellular selection; we now need to check how it translates at time. Notice, along the front curve, a favorable gene will spread with time, perhaps a low spread at the start, but then increasing with time. By contrast, in the rear curve, the gene spread is 'fixed' by how widely it spreads by its gene conservation.

If we then leave ancient life behind, to jump to modern life we have a strange phenomenon, where both a desire to conserve sequence across life and the need to mutate, are expressed by genes. We cannot take a need of fatty acids to conserve sequence, or a need of RNA to conserve sequence, and say it is modern genes exactly, but we can use a trick of mathematics to express the intent. If we assume that the ancient need of evolution to express conservation is now along an 'imaginary' axis and here the genes as DNA try to conserve sequence, the trick will work. It will combine with a need of modern genes to mutate to express variety along a real axis. Both forms work together.

If anything, we can measure the effect in modern genes. The most conserved genes are ubiquitin, followed by the H4 histone, then H3 and down a scale, with hox genes in animals, or cytochrome c across parts of life. The conservation of these genes is high, against the more mutable genes, across smaller segments of life. In the new theory, the more highly conserved a gene appears across all populations, then, broadly, the wider has that gene spread across life.

Rather, while many evolutionists were startled by the non-Darwinian or *neutral* results, no evolutionists, confined to their area of study, dared to claim that the experts are wrong. If you are specialized in fossils, you stick to that. If you are specialized in the cause of cancer, that too is your study. You cannot go against experts in molecular theory, because the math backs them up. Apart from biologists, if the mathematicians said that the mathematics of molecular theory was incomplete, and that genes can

[32]Some experts argue that junk or parasitic DNA is ancient. I doubt this. Organisms filled with junk DNA are mostly higher plants and animals that evolved in the last half billion years. Prokaryote genes have only small amounts of junk, whereas eukarya genes can have 90% of DNA as junk. Ancient species, such as lungfish, accumulated junk over time, or in modern species, such as fruit flies, certain junk has multiplied in modern times.

[33] Junk or parasitic DNA might be latecomers on the evolutionary scene; it evolve because of complex replication systems where "cutting-room floor" DNA takes a life of its own. Still, we must be careful labeling DNA 'junk' because we do not know what it does.

develop for two pathways, then the theory can change, but no one will admit to that either.

One reason for an interest with the *neutral* theory, then, is that around the *neutral* point, genes are able to migrate wider than one species, even for latter evolved genes that came after.[34] Here, molecular conservation was first, and it applies over life, such that the more conserved a gene, the wider it has spread. By contrast, the *neutral* theory teaches that molecular evolution is 'non-Darwinian' or *neutral*, but only by ignoring effects of wide gene spread over life. The new model emphasizes the *neutral* rate, by assigning it a place, half way along the back curve. If you apply the theory that $\mu_k = \bar{\mu}$(average), then $\varepsilon_k = 1$:

$$X_k = \varepsilon_k/(1 + \varepsilon_k), \text{ then}$$
$$X_k = 0.5, \text{ without any fitness.}$$

With sex, say, any gene that has a gain above the *neutral* level (it mutates more slowly than the *neutral* rate) will enjoy a gain of frequency, even if cellular gain along the real axis falls. On the other hand, the *neutral* view of gain for the cellular axis mostly falls, or is *neutral* for cellular change, is still correct. The mathematics of how a molecular clock for the *neutral* range is also correct. Just, for large changes over huge scales, the existing *neutral* model is incorrect. Instead, experts should estimate the length that a DNA or protein existed by the conservation of sequence

Genes such as ubiquitin or histones were part of higher life. Every gene that evolved since to higher life, had ubiquitin or histones included (with a small exception). It starts with ubiquitin, and histones, H4, then H3, and then cytochrome c, then Hox genes in animals, and so on. For example, there are endless mutations in higher life, more than lower life. Again, assume the first histone was H4. If it mutates to form a new gene, it will lose frequency, and an advantage to spread. So instead, H4 mutates to H3, which then mutates, and so on down a scale.

In the 'politics' of evolution, too, the advocates of the *neutral* or non-Darwinian model of molecular evolution are regarded as heroes. They are clever, and experts at mathematics. They also stood against the Neo-Darwinist school of evolution, to prove how molecular selection can be neutral or non-Darwinian. Instead, the problem is that they used a wrong theory. They tested molecular selection against a *cellular* theory, so results went down, when they should have gone up. Here, molecular selection has a pre-Darwinian form of selection. It is embarrassing to tell experts that they made an error in testing molecular selection, yet they must correct it. This is crucial to correct the *neutral* theory.

---

[34] Note, any function can split along real and imaginary pathways. It applies in physics and engineering, just no one has though to apply it in biology to time itself.

In summary, the new model of molecular selection can work, with an adjustment. One cannot claim that life evolves with molecular selection disconnected from evolution, but instead, it is connected, but via a 'j' (or 'i') in the equations. The extra action is through evolution, but it started 500 to 100 million years before life as Darwinian selection. Once the two forms of selection are combined and recognized, many other paradoxes of evolution can be solved this way.

How life started set the mathematical model for how life develops, millions or billions of years into modern times. From the start, molecular selection laid the foundation of cellular selection. Historically though, the mathematics of cellular selection was solved first, but that history cannot alter how genes evolved as real populations. Life cannot have started with molecular selection first and then suddenly switched, to cellular gain leading, with no mechanism to suggest how. Instead, if life began with molecular selection, this would still be in place, even if far slower, billions of years later, in how it works.

Life evolves from not one, but two pathways of selection. While modern genes respond to fitness, which is expressed by genes mutating into fresh varieties, this is a modern effect. The first effect, from prelife, is for evolution to express a drive to complexity. This began with pre-genes, such as RNA, but it also with fatty acids, finding ways to express complexity, by seeking to conserve sequence.

## *2.7 Multi-Cellular Evolution*

If you trace the history of how life evolved, from prelife until modern humans, part of it is incomplete. It is the evolution of multicellular life. We know approximately, when it occurred. One billion years ago, there was evolution for multicellular organisms, that included algae, but not for animals. In Wikipedia, multicellular life evolved, maybe 46 times, from many causes, but unless it includes animals, we, humans, will not evolve. We must tie animal evolution to development of the hox genes, which led to fungi, choanoflagellates, and animals.

It seems there is now a solution.

In Wikipedia, there are different scenarios for multicellular life, but they do not tie, precisely, to the evolution of hox genes. Instead, a group researching for eLife has a scenario where the Colonial Theory (from Haeckel in 1874) modified for Metazoa (animals). Here, evolution of multicellular organisms now fits with how the rest of life evolved.

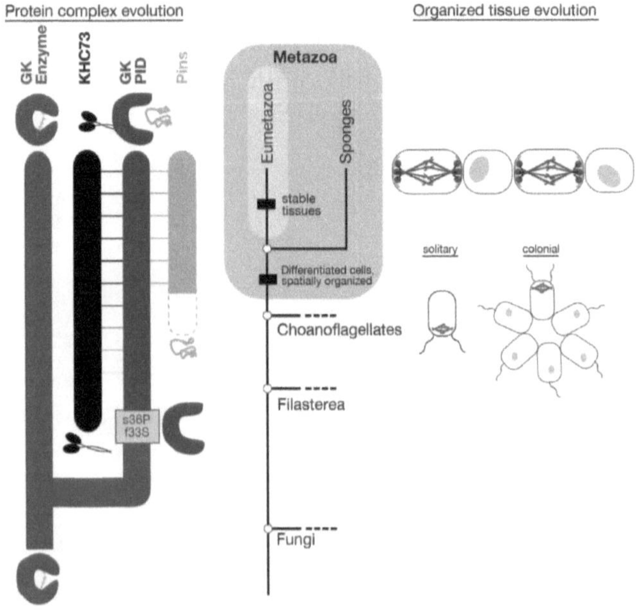


Fig 2.7.1 Douglas P Anderson and others, with a copy from eLife, for the evolution of multicellular theory in animals. This fits in with the other theory of the evolution of hox genes, and other effects in the evolution of these.

Next, consider the evolution of multicellular organisms.

If we look at the theory of life, there must be a progression. It goes from nonlife, into simple cells, such as archaea and bacteria. Then there is a progression from simple life into the evolution of eukarya. This was first just as mitosis, but later as meiosis, and then sex. Then was then a next phase, from single to multicellular organisms. This has reputedly occurred many times, and might have occurred even in bacteria. The problem here is that there must also be the evolution of animals. For instance, animals have 100–150 cell types, compared to 10–20 in plants, fungi, and protoctists. If we discovered another planet, the presence of life would be fascinating. However, unless life had evolved until animals, or the equivalent, this would not reveal the full effects. Similarly, unless we can explain the evolution of animals from all that existed previously, there is no conclusion that it leads to higher forms.

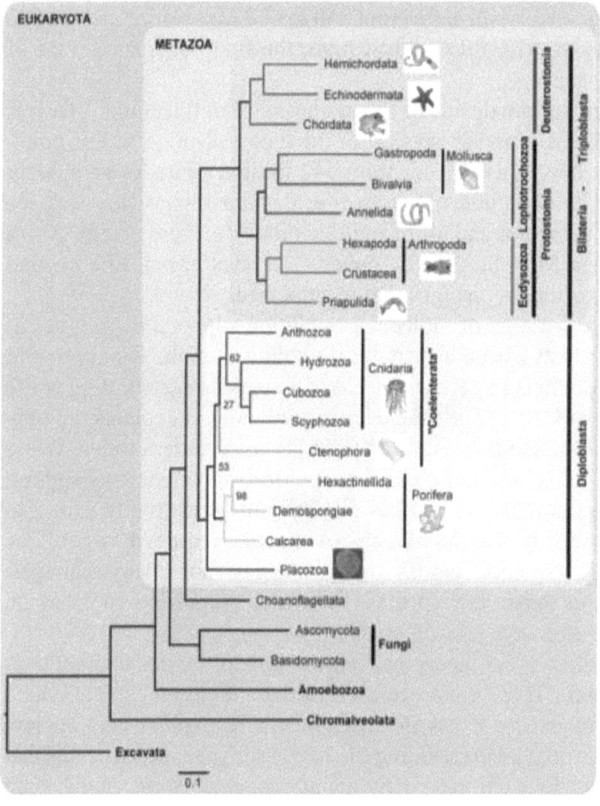


Fig 2.7.2 A scheme (from Wikipedia) on the evolution of life, including fungi, and other effects pre-Metazoan.

From the previous two diagrams, we can sort out the sequence of how animals evolved. Both diagrams assert that choanoflagellates were the last non-metazoan before conversion to animals. Fig 1.7.1 confirms that only animals evolved the GK-PID modifications that gave the evolution of that type. The claim that there were two types of modifications for animals, say, does not make sense.

We can now tie the evolution of the GK-PID modifications to the evolution of hox genes. The first hox genes originated early, perhaps in bacteria, but there noticeable spread was for the evolution of advanced eukarya, at least from fungi upwards. This was from about 1 billion years (or less) ago. Earlier organisms on the tree, including all the single celled varieties, go back towards the evolution of sex, at 1.2 billion years ago. We can now see a pattern, of sex at 1.2 billion years, then all the single celled organisms, then maybe fungi about a billion years ago, then animals evolved, as metazoans, at from about 600 Ma or slightly before that. Notice, there was a snowball earth from 850 to 630 Ma. Notice also, at 542 Ma, shortly after the GK-PID modification, the first punctuated rates of the Ediacaran began.

Now, after giving the details of the evolution of multicellular life, it is time to close off this introductory part of the theory. Notice, at this point, all the details of how life evolved, from 542 million years forward, were now in place. The evolution of lower life, then of higher life, sex, the neutral molecular theory, and all of multicellular life, were extant. All of this was before, at 542 Ma, the first punctuated rates began. Life beyond this was an adaptation, of changes that went before.

Notice, the new model of molecular evolution supports these types of changes. It has always been a dispute in evolution, if the changes were from random mutations in genes, or if these were deep seated, to how life began. The new model supports the deep-seated view. It explains not only how life began, but also how these effects play into modern times. While not an alliance, the new theory and the Evo-devo developments support each other. If the new theory teaches that ancient genes firstly, strive to conserve sequence, it follows developments of Evo-devo, where the conservation of sequence across many species, for many changes. Mutations, say, to make new species, lie in the supporting switches on either side of the hox genes, or other effects.

Notice too, there is no theory needed to explain the developments of the newer changes. If the theory relies on fitness alone to explain why a gene gains in frequency, it has no mechanisms to explain why ancient genes, such as the hox, keep repeating themselves. Especially, it seems that now the hox gene does not make new mutations, most of the changes are to gene switches, and these turn available cells on or off. A theory of genes gaining from favorable mutations alone is not complete.

This too, is why it is so important to sort out the sequence.

Although copied mostly from Wikipedia, the sequence of how the genes tie together is not presented clearly. If we follow the rule that a decisive change in a gene occurs once, it does not seem how multicellular organisms such as algae could evolve, perhaps a billion years ago. Yet Metazoans, with a decisive change could evolve at 600 Ma. Discovery that the Colonial Theory evolved once, for Metazoans, can put the theory in perspective. Certain multicellular organisms might have evolved before the last ice age, at 850 Ma, but metazoans (animals) did not evolve until the end of that, at 630 Ma. The pattern of life's development now fits. Notice, the punctuated radiations, stating with the Ediacaran, began only after the evolution of animals was in place.

This also places the next part, on punctuated changes in perspective. If we consider only the punctuated changes, for 542 Ma to the present, there are many variations, to be explained. However, if we reconsider that life evolved for near 4 billion years, the latter, punctuated parts were only 542 million years, in the last section. We still cannot state exactly why life took so long to evolve. Maybe it was the oxygen levels. Maybe it was carbon. Perhaps it was the salinity of the seawater. Maybe it was all the effects of the early snowball earths. We can state however, that biologically, the punctuated rates did not start until the preconditions, including the evolution of animals, were extant.

## 2.8 Fitness for Cellular Selection

The original punctuated rates of evolution, had it "post-542 million" years. (A post-542 million years is the modern era, when large changes evolved.) Effects for a punctuated rate of change, genetically, at least, were "pre-542 million" years. Punctuated rates need to be rethought, if these were the cause of the changes. Notice, this is not a theory, but the fact of how animal evolution occurred. Yet, can this solve the punctuated rates, any more than the existing theory?

Well, one effect is layers, between the evolution of lower life, archaea and bacteria, and the evolution of eukarya. Everyone should know this, but Andrew Knoll has given specific examples. For instance, eukarya can reproduce three ways, *aerobic*, then for *anaerobic* in a few species, and *photosynthesis* by algae and plants. Pre-eukarya species evolved for other cycles, however, such as processing nitrogen, sulfide, and even $CO_2$ gas, to produce methane. This plays a role in key processes, such as the decay of organisms. Other components, of fitness, metabolic adaptations, and punctuated rates, are how modern eukarya adapt. This was over the 2 to 3 billion years that lower life evolved. A final effect is the rise of oxygen. Multi-cellular organisms did not evolve, as recorded, while oxygen levels were low. Regardless of other effects, only the rise in oxygen, allowed multi-cellular organisms to increase.

When Darwin proposed that evolution was a struggle, it was assumed that creatures struggled to be more "like us". In truth, although simple organisms evolved first, this type is the most enduring. Bacterial designs billions of years old persist today and are prolific. Moreover, any time that simple life can refine incrementally, there is no impetus to increase in genome size or complexity.[35] Organisms are optimized to an existing struggle, so the 'fittest' does not infer the 'most complex'. No organism would enhance its fitness from becoming more complex, if a rival could become fitter from a smaller change. [36]

Again, the first point is that the "punctuated rates" of change were mostly over the pre-542 million years.[37] For example, the paradox of sex has been known for over 80 years, yet everyone thought that otherwise, evolution was fine. Then the punctuated rates of change were identified. People now thought that perhaps *molecular* selection was playing a role, but the model of it was the *cellular* change for fitness among the species.

---

[35] Individuals do not accumulate genetic complexity in single lifetimes. If among a group of individuals, those born with attributes that are more complex are reproductively favored in the struggle for life, over time life will evolve into forms that are more complex.

[36] Cyanobacterium is a well-known example of a simple organism over three billion years old, which has barely altered over that period.

[37] Term 'punctuated' is from Steven Gould and others, who also (famously) used the terms for the basis of equilibrium. Other new terms are my own.

When this was applied, rates of change became worse. Not only was the molecular change not Darwinian, but the theories proved that often it was *neutral*, or even non-Darwinian.

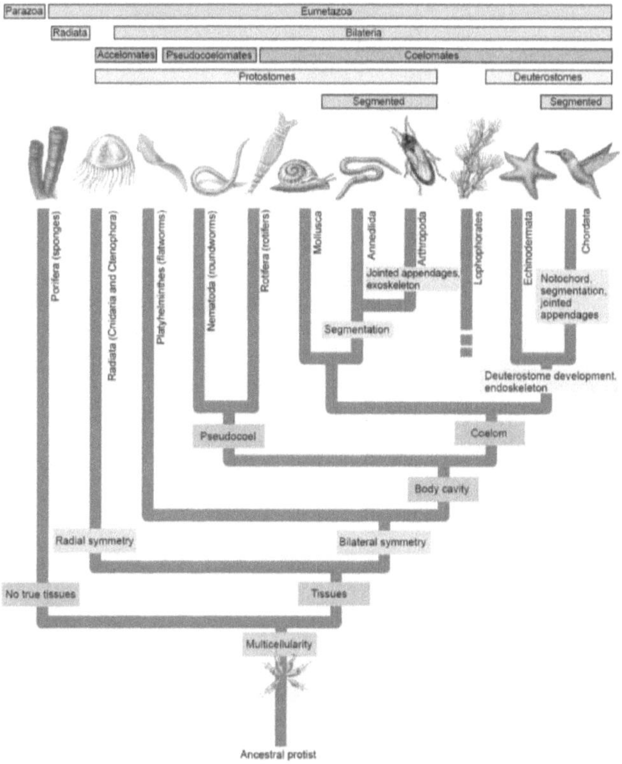


Fig 2.8.1 The theory of punctuated rates claimed that species changed little over long times. Much of this is simply the fact of how life evolved.

The next point is that I am the only person to connect the punctuated rates to the mathematics, but also think of it. If the rates of evolution are punctuated, but the only theory is of *cellular* selection, putting the two as one, will not work. A majority think that the mathematics of one pathway works, apart from reconciliation about sex. A minority think that the problem is not solved, but that the answer is cooperation among varieties. Except, neither view has solved the mathematics. If you take this into the evolutionist-creationist debates, neither view fully works.

Take the evolution of species over the recent area, as in Fig 2.8.1 previously. Now, anything copied from Wikipedia has disputes, such as did the rates increase that much, and so on. Yet something occurred 542 million years ago, and rates of evolution increased, along with extinction.

Over such time, the available places on Earth were filling up. Species also increase in metabolic complexity, and the evolvability of species went up. Species also became more intelligent over time.

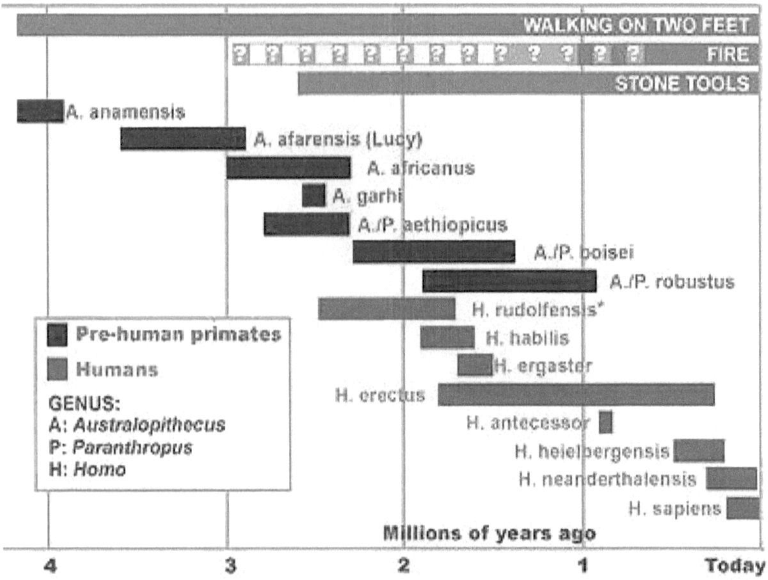


Fig 2.8.2 Typical arguments are the different rates at which species evolve. Here, rates are given for human evolution (copied from Wikipedia). There are many different reasons, for these varied rates. (It is the fact, but not the theory.)

Moreover, there are many reasons why fitness falls, but to see it, take an example from early life. Suppose this began in a hot, chemical-rich pool. After organisms adapted to a level of fitness for those conditions, less fit organisms were forced to migrate to another pool, perhaps cooler in temperature or less rich in chemical nutrients.[38] Small enhancements might give a rise in fitness, but eventually there would be a break, when the organisms must speciate.[39] At that, fitness for mutating genes falls, because although the new organisms become more adaptable, they must compete in less optimal conditions. They would pay a penalty, such as reduced offspring or longer reproductive cycles.

---

[38] Refer to early sections, about life evolving in the first forms.

[39] Fit creatures reproduce against low thermodynamic odds, so it is easy to make many copies of them. By contrast, complex creatures exist against high thermodynamic odds. It takes trillions of mutation-selection events to evolve land-living, endothermic, intelligent beings, such as a human. It takes far less selection-mutation events, against far less odds against it occurring at all, to reproduce single cells in a hot, chemical rich, fluid.

The next argument for fitness falling concerns how humans measure it. Verbal definitions of fitness result in circular logic (the fit survive and survivors are fit). Such tautology can be avoided by defining fitness as the DNA that an individual passes into subsequent generations. If we see a male bird with more chicks than other males, we consider him a 'fit' parent. If DNA tests reveal that he is not the true parent, then a rival male is 'fitter', and genes of the rival will become prevalent. If we applied this to the long-term landscape (if fitness is measured by passing on DNA) over time fitness must fall. For example, worms evolved before humans, so by definition the first worm had to be 'fitter' than any human. If anything, the first organism in life passed more DNA into subsequent life than any other organism.

While some genes can distribute widely across the biota, organisms cannot. The gene is not able to distribute across all life, across sea, land, and air, unless there has already evolved a broad food chain of plants, animals, and other types, such as lions that feed on zebra, and zebra that feed on grass. Genes in early eukarya, such as ubiquitin or the histones, gave rise to such chains, if a gene distributed widely in a lineage, fitness of that lineage would go up. Notice though, genes distribute widely not by one species, but by divergence in many species. As genes distribute across life, specialized organisms that genes utilize are left behind as new populations diverge and evolve.

For example, if a population speciates many genes from the original population spread into the new one (98.5% of genes in chimp ancestors spread in humans). Genes spread, but organisms that contain the genes do not spread into the new population, and now face a rival species that can displace it.[40] Moreover, genes do not spread in unison, but each gene maximizes opportunities. In the human-chimp example, 98.5% of genes gained advantage by distributing into the new species. The 1.5% of genes that were not incorporated into the new species were replaced or shed, so that these genes lost fitness via the speciation event. This effect applies across all the species. If anything, parts of evolution would be easier if it were against a loss of fitness.

One reason that the fitness landscape is undulating is that any quantity that increases without limit has runaway effects. By contrast, if evolution contained a countervailing force that acted to keep life simple, this would avoid the paradox. It would also better explain the layered structure of life, or unequal rates of evolution. The idea is that fitness rises, to a particular level, but then life strikes a barrier, which it can overcome by a loss of fitness before it can evolve to the next level.[41] In the undulating model,

[40] It is similar to how the ancestors of chimps provided the founding gene pool for human evolution, but modern chimps now face deadly competition from humans.

[41] There is a so-called *Red Queen* effect, where fitness is a cyclic adaptation. In bacterial cultures say, one generation does not breed 'fitter', but adapts about a mean.

fitness at a locus climbs up to a peak. There are higher peaks that the gene could attain, but the gene must descend against a loss of fitness, before it could ascend higher.

Moreover, despite that it seems odd that fitness falls, examples of it occur. Sex is seen as a 50% loss of fitness against pre-sexual production. Not just for sex, but also higher behaviors, such as love or cooperation, are hard to explain as a gain of fitness against easier means to pass on DNA[42]. For instance, while simple creatures reproduce as much as they can, life forms such as mammals or birds choose mates carefully, and reproduce by behaviors that limit DNA passed on. Bacteria produce in greater numbers, at a far less cost of energy, at less time to reproduce, in a less delicate web of life, than complex creatures. If scientists look for first life, or life on an inhospitable world, they look for life that can survive at low costs against the odds of it being there.

In this, a macro-mutation might appear as a step change, for example, the mutation that added a second atrium to the heart. Over the history of life though, a mutation resulting in an extra atrium has proven only fit if there exists a fish that has a lung, but does not have a second atrium. In the history of life, the probability of that event was once. (There are two probability events; one for a fish having a lung but no second atrium, and one for second atrium. Yet the probability of both events occurring is low.)[43] Even if it occurred, it would never be as fit against a rival that had already matured with that novelty. A hox gene set can duplicate once then once more, but never duplicate again after that.

Perhaps, far in the future this will change. Perhaps we will discover life on other planets. Perhaps we will discover if the evolution of life on earth was fast or slow. Perhaps we will discover the chances of advanced life evolving. For now though, this is speculation. I think that the chances of primitive life evolving, to lower life, are quite good, but the possibility of advanced life, even to intelligent creatures, have not been discovered as yet. In the universe, the earth is relatively young. To imagine advanced life, within a practicable distance, for now seems remote. It could change, but we must deal with facts that we know. These are the evolution of humans, only on earth so far.

Now, consider the model of adaptation and speciation.

Again, in the standard theory, adaption occurs at one locus at a time. Any favored, single gene or allele will spread, and it will drive up mean fitness of the population about that locus. Note though, that once a gene becomes 100% distributed in an original population, it is assumed as

---

[42] In sexual reproduction, 99% of genes in an individual pass on by sequence, just that the genome is reshuffled each generation. Via sex, the individual gene gains in $\varepsilon_i$ what it loses in $x_i$. However, the exact copy of the entire genome goes down

[43] This would require an equation such that if an improbable event occurred even once, it could not happen again. I am not sure which functions, if any, work that way.

'fixed'. Once a gene is 100% distributed it can gain no further fitness, because 100% distribution is already the maximum.

Still, a gene distributed 100% in a population could distribute wider if the original population divided into new species. Except that in standard models adaptation about a locus, and the evolution of a new species is the same process, but accumulated over longer periods. A gene, $x_1$, adapts about locus $X$. When that is complete a gene $y_1$, adapts at locus $Y$. The cumulative effect of these adaptations, $X, Y, Z$ ... over many loci results in a new species. This is how Darwin might have seen it. This is called *anagenesis*. These small changes one locus at a time occur until enough changes accumulate that a new species replaces the previous one.

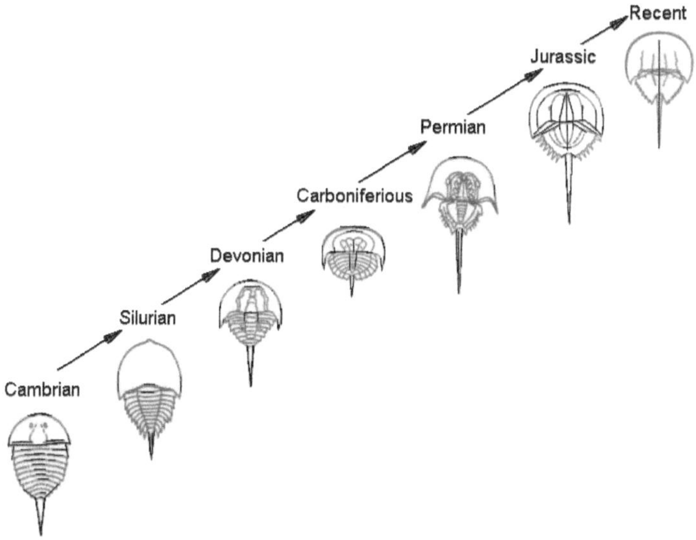


Fig 2.8.3 Evolution by *Anagenesis* for horseshoe crabs, over 500 Ma. These are small adaptive changes over time, until enough changes accumulate so that one species evolves into the next. Yet even for horseshoe crabs, there is dispute if this really is *anagenesis*, or there are branching species missing from the fossil record. (Redrawn from Strickberger and adapted from Newel.)

However, there are some well-known difficulties with this model. The first (that Darwin could not have known of) is that each time a new gene spreads, $x_1$, $x_2$, $x_3$, the mean fitness of the population must rise to spread the gene. Because it is in the same population, the mean fitness of that population is increasing.

Evolution though, does not consistently unfold via *anagenesis*. Small improvements to species over time are observed, such as for horseshoe crabs shown in Fig 2.8.3, or for evolution of the shark shown in Fig 2.11.1. However, evolution of new species seems to occur more rapidly. For example, over the period that horseshoe crabs or sharks improved slightly, dinosaurs rose and fell, and the Earth was repopulated by new orders such as birds, ungulates, or primates. [44]

The other problem with the *anagenesis* model at first seems trivial. It is that it is hard to define at which incremental change in an old species will produce a new one. Creationists now concede that species can adapt one locus at a time, so evolutionists now need more compelling evidence as proof of speciation than that. However, the scientist also needs to quantify at which point a new species has evolved, and this favors a model of speciation called *cladogenesis*. One population subdivides for another for many reasons[45]. As the populations drift apart genetically, until a point where the members can no longer interbreed, then this is the point where speciation can be defined. [46]

One insight is why evolution might favor *cladogenesis* for speciation, because *anagenesis* is of limited benefit to a gene. In anagenesis, a gene spreads 100% in its population, then another gene alters, and so on, and the population evolves over time. Yet this is one population. If extinction strikes, a single population can terminate, so the gene loses distribution anyway. Random extinction might have killed the dinosaurs, but if the histones, ubiquitin, and hox had distributed into mammals its sequence would survive. Modern genes that have altered many times are less costly. Junk DNA that does not code any can alter at no cost other than its perpetuation (but see footnote[47]).

Again, Sean Carroll has modified this in his latest book. (Note that the synonymous sites are where changes of DNA do not affect fitness.)

> This pattern of the strong preservation of the protein sequence at most sites, with the synonymous evolution of the corresponding DNA sequence, and diversity limited to a few sites in the protein, is the predominant pattern of evolution of the DNA record.

In summary, a challenge of large-scale evolution is not just to explain why new populations evolve from a founding one. It is how continuous, steady change produces the punctuated pattern in the fossil record. This is not just in the layered geology of life, or that after long periods of stasis

---

[44]Many people do not see this as a problem. A million years is fast in the geological record. Yet, can anyone conceive how the populations of Earth might look, in a million years time?

[45] Again, creationists now seem to accept adaptation within a broad type, such as birds. My view of this is locus-at-a-time adaptation, rather than transformation of the genome.

[46] There is plenty of evidence of speciation in modern times. New species evolved in the plant genus Tragopogon in 50-60 years, or modern wheat evolved from a wild type, within the last few thousand years. New snake species adapted to ocean life including live birth in a reptile in modern times. Many plant, fish and insects have also split into separate species.

[47] The neutral theory says that there is no cost to alter DNA; it just happens. Again though, this assumes that conservation rates have no effect on distribution. By contrast, this theory says that costs are proportionate to the degree of gene conservation. Altering a conserved gene is a high cost, altering junk or unexpressed DNA is a neutral cost.

new species appear to evolve. Instead, as genes radiate outwards into new populations, other genes in the genomes must alter, but in doing so, they would lose fitness against drastic change.

The next problem was the alteration in species between *anagenesis* and *cladogenesis*. Still, the new model can show why there is a steady genetic pressure, not just from organisms, but also for highly conserved genes to increase distribution across life. This need of genes to conserve sequence can better explain other effects that produce new populations. This too can help clarify, how adaptation within a single population, is a different process in its effect on gene distribution, than speciation across many populations. Referring again to the hox genes, we can see why they stayed nearly the same and highly conserved, over millions of new species to evolve as life modified.

The effects of genes spreading from early life to gain distribution, while inducing weaker genes to alter to adapt, with the process enacting against geological change, help produce the large-scale effects.

## 2.9 The Mechanisms of Change

When scientists discovered that light elements such as hydrogen and helium could burn as nuclear fuels in the interiors of stars, fusing into heavy elements and releasing energy, it seemed that they had discovered how the heavy elements formed. However, the fusion of light elements only continues until the element 26, iron. From here, fusion becomes energy absorbing (not releasing), so it occurs against a *loss* (not gain) of energy. This required a much "bigger furnace" than the normal burning of stars. Eventually, it was discovered that a supernova could produce the energy needed to form the heavier elements.

This is something of the dilemma here.

As long as we thought that new species arose steadily, there was no need for a special radiation to bring changes. However, the change is complex, and there is a rise of fitness along the axis of gene conservation, but a fall of fitness as species seek minor adjustments. In this case, there must be some special type of explosive event, to make evolution into new types in a radiation viable.

Now in the history of life, explosions of new classes and phyla have occurred analogous to a supernova. Except the astrophysical supernovae, have calculable causes, based on the star's mass. By contrast, explosive radiations of life on Earth seem to have unique causes in each case. For example, the evolution of organisms that were able to photosynthesize solar energy, converting carbon to oxygen, led to an explosion of types able to exploit this energy source. In turn, a buildup of oxygen in the atmosphere released an explosion of aerobic types, able to adapt at higher metabolic levels. A cooling of the climate (and another oxygen increase) at the end of the Cretaceous period released a new explosion of types that had pre-adapted for endothermic regulation, such as mammals and birds. Instances of a build-up, followed by a release that appears explosive, do occur. Despite this, an adaptive radiation might not be a 'supernova' that overcomes the fitness loss.

Instead, this chapter will suggest that adaptive radiations, while they appear explosive, might not put pressure on species to alter radically. In a radiation, new species spread rapidly into newly opened ranges, but the evolution of types leading to radiation often occurs slowly. The radiation of mammals 65 Ma ago seemed explosive. However, mammals radiated because high cost changes were paid in the previous 140 Ma as mammals accumulated novelties such as body hair. The Cambrian Explosion was also the result of an accumulation of novelty that might have begun with previous evolution, half a billion years earlier. Types radiating into open or vacated ranges, might also not be undergoing genuine transformation. Radiating types might instead be adapting established designs, while also avoiding radical experimentation.

Instead of a radiation, this book will propose that the big 'furnace' of evolutionary change is not an explosive event, but a pressure across all biota that pushes some populations into peripheral *niches*, on the fringes of flourishing biota. Peripheral niches are harsh or outlying ranges, which are difficult to penetrate for a first time. [48]

For example, the oceans are huge, but if enough variety of water-life evolves, eventually all the ocean niches will be filled. Creatures can then penetrate shorelines, and then onto land. However, to penetrate onto land requires innovations (such as lungs, limbs, and a non-dehydrating skin) and large changes to genomes, which will consume fitness. In this way, amphibians evolved in a shoreline *niche*, which was peripheral to oceans, where life evolved submerged. Flowering plants likely evolved on the edges of swamps, as plants began to migrate onto dry land. Birds perhaps evolved in a niche of trees, high peaks, or colder habitats. Life's largest transition, from prokaryotic to eukaryotic life, has no niche hypothesis of how it evolved. However, eukarya evolved from a symbiotic forcing of lower life together, under extra force.

Consider the modern sequence. There was an era of intense glaciations (850-635 Ma ago) with a 'snowball Earth', which would cause a constriction of life. Then, body material such as tissue, nerves, and blood evolved, which was after body segmentation. Then came the Ediacaran (635-542 Ma ago), when multi-cellular organisms evolved. However, the key change was the evolution of sex in animals, about 630 Ma. After that the Cambrian occurred, and the origin of many types in a short time (542-480 Ma ago). This was after the Epicedian and sex for animals. When the Cambrian occurred, it was designs encountering new conditions, such as increased oxygen levels.

Another radiation was the proliferation of mammals and birds was the Cretaceous-Paleogene (or Cretaceous-Tertiary). Again, the radiation was sudden, from 65-55 Ma ago, with new orders evolving in a short period. Yet the experimentation began 200 Ma earlier, following the Permian radiation.[49] Lineages giving dinosaurs, modern reptiles, and mammals, branched then. Flowering plants evolved since the Permian, but widely radiated near the end of the Cretaceous.[50]

Note too, that while the evolution of dinosaurs and mammals began at a similar point, it followed different paths; during the Permian (300-250 Ma ago) proto-mammals radiated widely. After the Permian extinction,

---

[48] Though it occurs with human populations, it does not always follow that because there is overcrowding, populations are pushed outwards towards a periphery of existence. However, another effect called saturation, discussed later, will cause migration.

[49] This is simplistic. There were three branches; anapsida, synapsida, diapsida, and a fourth, euryapsida. Dinosaurs, snakes, lizards, birds and crocodiles are diapsida. Modern turtles are anapsida. Mammals evolved from synapsida. The fourth line went extinct.

[50] The radiation of flowering plants was also interactive with the radiation of insects.

dinosaurs underwent wider radiation, and flourished into thousands of species. However, during dinosaur radiation the line leading to mammals suffered constriction, at times into a small, nocturnal, species. Despite this, during a period of dinosaur radiation, dinosaurs did not experiment with radical new attributes. By contrast, over the same period, mammals innovated with hair, mammary glands, live birth, endothermic regulation, enhanced dentition, a diaphragm, and many other novelties.[51] Notice, once mammals radiated it was over a short time window of 10 Ma (65-55 Ma ago), compared with the much longer 140 Ma of the experimental phase of mammal evolution.

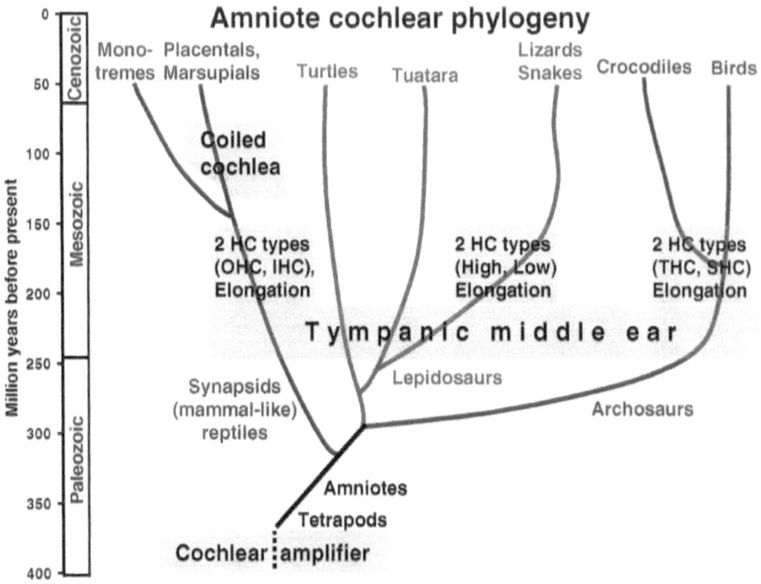


Fig 2.9.1 Mammals did not evolve from dinosaurs, but evolved in parallel from a primitive reptilian ancestor. (Note, this is pasted from Wikipedia, against a simpler diagram that I used in earlier books.)

Why is a long development phase needed for major innovation?

Well, it is easy to explain why a long time is needed, except that the developing lineage should not radiate too early. Constriction confines a lineage within a peripheral niche, where experimentation will have more

[51] There is debate over why mammal innovation was not as well preserved as the Cambrian radiation. Evolution of key mammalian innovations, milk, fur, live births, and four-chamber heart, were soft modifications, not easily preserved as skeletal structures. Cambrian fossils were also captured in one rich formation, but mammal evolution was a constricted range of many small populations, over the entire Jurassic-Cretaceous period.

opportunity to mature. Harsher conditions are also likely to speed up rates of mutation. By contrast, in an adaptive radiation, species would diverge into opened ranges, under less pressure of competition and stress. To give some idea on mutation rates, it took 140 Ma (from 208-65 Ma ago) for mammals to evolve. There are some 30,000 genes in a mammal genome, of which about half might be involved in alteration. One gene altering every 10,000 years is a fast rate, but not an impossible one. Suppose the average rate of alteration per gene was $10^{-8}$. In that case, even if mutation accelerated one thousand to a million times faster, it will still not be more than one or two changes every few generations. Mammals also endured plateaus of stasis over millions of years. There were long periods of normal mutation rates, punctuated by rapid change in shorter periods.

Moreover, although radiation can result in a many new species, the type of change during a radiation might not be hard to adapt.

Newly penetrated ranges into which less fit types might be pushed, allows species to innovate in ways that might be fatal in an open range. In the age of dinosaurs, predators and prey competed to run faster, but none of them evolved endothermic regulation, body hair, or a fully sealed four-chamber heart. Any dinosaur could have run faster for longer with these attributes, but there was no chance to evolve them in the daylight range. Instead, it was only when proto-mammals entered the nocturnal niche, which cold-blooded reptiles could not penetrate that the attributes had a chance to evolve.

Moreover, even minor speciation requires an isolating or peripheral range for it to occur.[52] Usually, geological isolation must first divide the population so that individuals from the divided populations can no longer interbreed. Even if there were no isolating boundary, within a porous range overpopulation could force part of the population to peel off, to go its separate way. Even this would bring pressure on the outer population as an isolating barrier that acted as fitness.

Consider, say, a collection of individuals, bound to each other by a shared attribute. Mostly this is shared reproduction, so that the outer rim of where reproductive partners can be found is the geographic range of the species. Outside of this range, any individual would not have a reproductive mate, but other factors lead to the outer range becoming peripheral. If the species has survived many generations it must be a proven design, and within that design individuals can enjoy high fitness for small changes in genotype. Say, if a male was born with an advantage that was a small deviation from the species, the male could mate with any of the females to enjoy high fitness. However, while every species is bound to a successful design, it is also a design optimized around a particular spot

---

[52] Speciation among polyploidy plants or by non-sexual gene exchange might not require forcing. It would be interesting if such speciation involved any measurable fitness loss.

in the range. If it is optimized for life in the deep forest, members can survive at the edge of the forest, but not in optimum conditions. So, although new novelties can evolve anywhere in the forest, the deep forest design will predominate.

To move out from the forest, if there were an ecological change to the forest, might require large changes in the inhabitants, such as change of locomotion. For stable conditions, fitness will accrue fastest for easy-to-enact changes (bigger, stronger, sharper, etc.) It has a retention effect, because fitness-rich adaptations will favor a deep forest design, because the easiest changes likely work best in the deep forest.[53]

Similarly, we can go to how genes increased the scale of information also by doubling. Now, apart from the "evolution-creation" controversy, we have a "scale of order" controversy, also used by scientists. Before evolution, scientists had a "scale or order", and modern scientists want to disprove it, by making humans a similar scale to lower animals, and so on. Yet there is a strict metabolic scale of order, in that milk-producing animals are on a higher scale than non-milk animals, so for the record, mammals are higher than fish, while humans might, for many reasons, be highest of all. This can be worked out, by increased metabolic rates of advanced organisms, from doubling, and many other increases in gene conservation, as life evolved.

Any environmental range, that a population is forced to adapt outside of its norm is a *peripheral* niche. Plains are peripheral to forest dwellers. Cold is peripheral to thermophilic life. The shoreline is peripheral to fish. Night is peripheral to reptiles. An unpleasant or low quality diet is peripheral for minor speciation. No individual wants to enter these, but when there is great pressure acting across all of life, some types are forced into peripheral niches to escape fierce competition in crowded ranges. Isolated and protected from other species, a line of organisms can afford experiments that incur fitness losses. These losses might not be viable in a more ordered range.

To give an analogy, consider how the wind blows steadily across an ocean and the sea tries to move with the wind. Although they are both fluid, a large mass of water is harder to move than a mass of air. As a result, near the surface it is easier for water to bob up and down, than to move in one direction. This creates an energy wave, which travels for hundreds of miles. Near the coast, the sea becomes shallow, so now it is easier to move water than land, and the waves start to break. The land will move too, but if it is a rocky coast, it will erode in steps. Continuous movement of air across the open sea is producing movement that causes step changes in a coastline hundreds of miles away.

---

[53] If there were a fitness gain, species would divide too readily. However, species keep high retention under stable conditions, despite constant change and mutation.

The action of wind on water would not be easy to visualize if we had not seen it. (How many people realize that waves form because it is easier to move wind than water?) However, we cannot picture evolution as easily as ocean waves. We only observe via the fossil record that new species evolve suddenly, and then persist unchanged for long periods. We also note that genes appear to mutate at steady rates, and that DNA might seem easy to modify. This way, alteration of DNA at a steady rate should produce evolution at steady rate. Small changes in evolution do occur steadily, but on the scale of an ocean to a droplet, evolution seems to act in step changes, which is much harder to explain.

This is why there are stepped changes, as a 'dam-burst'. (Rain falls and the river rises, but when the dam bursts, there is an upset.) However, the real dam is barriers. Nature will only evolve complexity if there is a cost-efficient way.

As existing biota saturate, populations are pushed into marginal niches. Reduced competition in a *niche* allows organisms a wider fitness cost margin. Any change to genes is costly, but there are many ways to pay the cost. Evolution from prokaryotic to eukaryotic life required a huge change (DNA in eukarya had to be reconfigured) which took a billion years. Once it evolved, though, eukarya had mechanisms that allowed genes and chromosomes to be copied first and modified later. This lowered the cost of change by providing DNA templates for future designs. Sex lowered the cost of change by providing greater variety at reduced risk of disruption. Although new genes had to evolve for eukarya, the cost was paid in a billion years of accumulation. By contrast, humans evolved from great apes in a few million years, but it required 1-2% of genes to be modified. This way the costs of human evolution were paid much earlier, in the evolution of mammals, or earlier still, in the evolution of sex or eukarya life.

If easy-to-change adaptations (length, shape, size) are available, individuals will compete to consume these first. Competing individuals struggling for any slight advantage cannot afford experimentation with complex novelties that require long times to perfect. Most times the fittest move is an adaptation like an existing attribute, such as a limb, and modifying its shape, size, length, or function. Even then, every species is locked into its own fitness *trajectory*. Any trait encoded in DNA can be easy-to-evolve, but they are only easy-to-evolve if there is a fitness need for them, in rivalry to other means of adaptation. This gives a stepped effect, because organisms try to adapt easy-to-alter attributes first, rather than incur a fitness penalty of changing too much.

Life evolves against a cost that correlates with the thermodynamic probability against the organism existing. If there is no measure of cost, any outcome of mutation and selection is likely. Yet in the new model, a gene's resistance to altering sequence quantifies the cost of changing it.

Over billions of reproductions, loss of sequence is an effective loss of fitness for the gene on the front axis, and loss of copy of the genome is a loss of fitness for the organism. Individuals who spread their type will always be fit, but over a *lineage* that evolves from a simpler ancestor to a complex descendent, cellular fitness falls. However, organisms still try to avoid a loss of copy. If a trait like a wing, eye or flipper is easy to evolve, it does many times.[54] Yet if a trait like limbs, feathers, or the four-chamber heart evolved at a fitness cost, no modern creature can afford the penalty to evolve that novelty again, facing the fiercer competition that the new novelty brings.

Except if various pressures force organisms to increase in complexity, the creatures increase in ability to adapt faster. Learned, social behavior can be adapted faster than inherited behavior. A hand can be adapted faster to a new means of food gathering than a paw. Flexible limbs and backbone can adapt to new means of locomotion faster than rigid designs. However, flexibility causes *saturation*. If primates can already adapt fastest of all species to environmental changes, phylogenies of slower adapting species become saturated due to like competition. If a cat has a paw, but the primate already has a hand, a cat cannot evolve to obtain a hand at the speed that a primate can adapt behavior, perhaps to grasp tools.

This leads to a circumstance when not just some, but all the available pathways become saturated.

[54] Wings evolved on pterosaurs, flying fish, birds, bats, and insects.

## 2.10 Easy and Hard Changes

This book depends for its validation on the hard work of science: This book depends for its validation on the hard work of science: research, data collection, and math. Many of the chapters contain novel ideas, but by the end of each chapter, there is nothing left to say, that does not depend first on more research. Insight can take one so far, but then it is time to take stock of what is known.

In preparing this book, though, there arose an overriding insight into how genes compete and evolution works that all evolutionists might wish to consider. Throughout these chapters we must keep asking why genes try to conserve sequence to such an extent. The answers are in molecular chemistry and selection; yet is that all? Some viral or parasitic fragments do not try to conserve sequence, but mutate opportunistically to spread in any way they can. So why do not all genes behave like this? Why should life not be a Malthusian soup of genes, without organization? A need of genes to conserve sequence drives the macro-structures of life, but it still does not explain why genes try to conserve sequence at all. Not that one could explain it in ultimate terms of why the universe exists. However, we can at least ask; how does the universe being the way it is, affect how genes distribute, and how life evolves?

Wondering this gives and insight. In a universe of many properties, *some properties are easier to alter than others are.*

We see this all around us.

In physics, it takes immense energy to alter the structure of matter, but small energy to bind existing matter into new forms. As a result, it is easier to rearrange elements as molecules, than transform one element into another. These properties give direction to chemical change, and this carries into life. Chemistry is the basis of life, but the elements of life; water, amino acids, RNA and DNA molecules, all exhibit strong, hard-to-alter bonds on one molecular axis, but weak, easy-to-alter bonds along another. The water molecule, $H_2O$ is a tight chemical bond between two $H^+$ atoms and one $O^{2-}$ atom. Yet the two hydrogen atoms sit slightly to one side of the molecule, so it becomes polarized (+) on the hydrogen side, and (–) on the oxygen side. Breaking the chemical bond is a hard change, but breaking the polarizing bond is an easy change. When life evolved, the easy or hard to alter physical and chemical properties of the universe became the backdrop against which organisms competed for each slight advantage.

For instance, organisms that altered easily changed attributes did so for less cost to resources, so their type spread. Types that found low cost ways to adapt competed again, for further refinement in adaptation. Yet entities that live also die, so a way must evolve to pass adaptations on in an incorruptible form.[55] Genes achieve this by encoding the protein structure of each attribute, but genes also encode, less precisely, another piece of information. This information is the viability of the attribute that the gene expresses and the cost to alter it. Genes that persisted through billions of selective events largely unaltered did so by expressing the attributes vital to each organism, which it would be fatal to change.[56] Genes that altered slightly into a family of types did so because the attributes that they expressed were adaptable to each species. By contrast, genes that continuously undergo mutation do so because they no longer express useful attributes, but they multiply by any means as junk or parasitic gene fragments.

Richard Dawkins has said that the need of early genes to replicate their pattern is the need of all entities to achieve stability. As explained, successful genes, conserved across the widest number of organisms, are inherently the most stable. Even so, R Dawkins is treating stability of the gene as the cause itself. However, DNA is not "life", but an information code, where sections of the code can alter each reproduction. Other parts of the code are 'hardened' as highly conserved genes, which barely altered sequence in billions of years.

Moreover, genes do not 'want' to retain copy fidelity, any more than they 'want' to alter to copy in great numbers. Genes exist in ways that pass successful adaptations to the next generation. The most successful genes are designs that are recyclable into a variety of types unaltered, which are expressed by sequences that are most stable, and so forth down a scale. When selection conserves some sequences, but allows others to alter, genes carry forward from earliest life a tenet of existence that some properties are always easier to alter than others are.

Not only that, but genes resist change because large mutations are often fatal.[57] This is why individuals that adapt from minor changes to core genes, adapt the easiest alterations first. This also is why evolution contains so much *homology*, or adaptation of basic body plans by shared ancestry. Limbs can be adapted into flippers, wing, legs, arms, or even atrophy as vestigial organs, like legs in snakes. In this sense, adapting limbs into new forms is an *easy* evolutionary change that occurred many times. By contrast, evolving the four-limb, vertebrate body plan was a *hard* to enact, episodic change that occurred once only, and is now fixed in present life on Earth.

---

[55] We presume that the first replicating cell passed on its entire self. Yet as cells became more complex, it became efficient just to pass on the code of how to replicate the cell.
[56] The effect is self-reinforcing. Mutations to crucial sequences abort in the fetus, which inherently conserves the sequences of genes that survive. Selectively too, it would not be fit to waste reproductive resources on mutations that will not be viable anyway.
[57] See Appendix I, but the 3rd letter in DNA can often alter without changing the amino acid. It would be more one change in the 1st or 2nd letter, or two changes in the 3rd letter that would cause a fatal disruption.

The reason that some changes would be harder to evolve than others are, is the statistical difficulty to evolve. Evolution does not violate the Second Law of Thermodynamics. Even black holes do not violate it, but evolution progresses against the direction of thermodynamic probability. The most probable flow of thermal energy is from a hot to a cold body (a hot meal in a cold room cools over time.) The most probable flow of organizational energy, is from an ordered to disordered state. (It is easy to shatter a glass by dropping it, but it is impossible to reassemble the shattered pieces into a new glass by the reverse process.) Life is a more ordered state than a disordered one, and it usually exists at a higher temperature than its surroundings. However, a high probability against life need not have been how life started.

Instead, suppose that first life composed a cooler bubble inside a hot broth. In that case, first life had to overcome an information gradient, but not a thermal gradient, which allows a higher probability of life starting. However, once life had started, each time it increased in thermodynamic complexity, it became thermodynamically less probable. Simple life at a lower thermal gradient than its surroundings is more probable than some people think. On the other hand, a human being metabolizing 250 million miles away on the cold, airless moon is an event so thermodynamically improbable that it could never have occurred by chance. Natural selection has a way to overcome steep improbability gradients, but this takes time. The more thermodynamically improbable the result, the longer it takes to achieve. That is why a complex organism is far less thermodynamically probable than a simple one. It is why the more complex organism takes longer to evolve.

In life too, just as some changes are harder to adapt than others are, some changes extract larger costs to implement. One cost is the energy involved in change. However, cost can be mitigated over long times by encoding the results of prior selection in genes. Genes accumulate the results of prior selection, similar to how humans accumulate the fruits of prior civilization, in money or knowledge.[58] For instance, eukarya life appeared after a billion years of prokaryotic accumulation. The Cambrian explosion was sudden, but only following a billion years of eukarya accumulation. Following the Cretaceous extinction, mammals radiated after 240 Ma of prior accumulation during the dinosaur era, in which many new attributes had time to evolve. This is why the evolution of new species, classes, or phyla, is an order-of-magnitude more difficult than

---

[58] Initially, wealth for humans tallied possessions or labor, and knowledge was passed on by custom. Once humans have money and writing, wealth and knowledge can take new forms, accumulate, and move from place to place. Today it would be too expensive to colonize the planets, but wealth and knowledge for such a project will accumulate, so what is prohibitive today will be possible tomorrow. Similarly, behind the industrialization of the last two centuries lay thousands of years of accumulation of human wealth and wisdom.

adaptation of existing species. There has been adaptive variation among orders of mammals in the last 55 Ma. However, the existing orders (carnivores, ungulates, etc.) evolved within a short period 65-55 Ma ago, and no new orders have evolved since. It is because adaptive variations on a body plan do not require radical changes to established designs. By contrast, the evolution of new body plans requires costly changes to existing genomes, which are hard to enact changes.

For long accumulation, the cost of change is high, but once a new design is perfected, new species can quickly adapt, because the high costs have already been paid. If an initial cost of alteration is high, though, no rival organism can overcome the cost to evolve that trait again. For instance, there was a high cost to evolve the vertebrate body plan. Yet once it evolved life adapted into thousands of vertebrate species. Except now, no creature not a vertebrate could overcome the fitness cost to evolve into a vertebrate, in rivalry with organisms already leveraging other advantages of the vertebrate body plan to adapt.

Evolution is the survival of the fittest, but there have been many evolutionary convulsions (asteroid strikes, and so on) in the history of the planet. In that case, all the organisms alive today descended from the varieties that survived the convulsions. They partly survived by mutating their way into new varieties. Mutation can be costly if there is no call, but it is a ray of hope if all else fails. Genes, or combinations of them, that knew when to mutate and when not to, survived and spread inside organisms that survived and spread for the same reason. Genes do not know that other populations are being driven extinct, but they sense when there is pressure on the one population where they are resident. Genes under pressure throw up mutations at a faster rate because that is how genes, and life, survive periodic extinction. If pressure goes off, less advantaged populations can migrate, rather than be forced into extinction, and the mutation rates will stabilize.

It is not intended to make a great issue about accelerated mutation rates. As mentioned elsewhere, genes such as heat shock speed up rates, and other genes such as hox genes find ways to induce large genomic changes. Other genes, such as histones, have no reason to alter rates, as they will spread anyway. Under pressure though, effects of mutation become pronounced. In stable times, species optimize to an environment, so even if new mutations are not harmful, they might not bring advantage either. Nevertheless, once conditions change bizarre or otherwise useless mutations can suddenly be advantageous. Introducing a pesticide "speeds up mutation", in a population under attack. Yet it could be a combination of the pesticide inducing mutations, genes reacting to stress, or mutated individuals spreading. The total sample shows only that mutation rate has increased, but it could be for several reasons.

To select which alterations will be fit for conditions, organisms need a measure of when a simple change works, or when a redesign is needed. Genes that express cell materials and functions will be hard to alter, but changes of size, shape, color, diet, or behavior will be easier.[59]

Every organism pays a price to adapt, which is calculated as a cost to alter core sequences, so over billions of years life adapted new ways to express variation without altering core sequences. Eukarya can express variation by duplicating and rearranging chromosomes or genes, at less cost to alter sequences than in bacteria. Sex lowers the cost further, by reshuffling genomes each reproduction. Over ages of fish, reptiles, and mammals, selection sought the least costly ways for organisms to adapt. Selection also refined to where life could adapt at a maximum rate. So that once primates evolved, the only way to adapt faster with less cost to altering genes was by culture and learning, which might explain human evolution in that context.

In summary, many physical properties of the universe are easier to alter than others are, and the effect carries into life. Genes encode designs in ways that can be passed from one generation to the next. However, genes also encode, less precisely, the "cost" or difficulty to perfect any design. This "cost" correlates as the resistance to alteration of any gene, such that key designs will be harder to alter from an original sequence for many reasons. Each gene will evolve a strategy for how it can express useful variation, but at the same time conserve copy. Moreover, any gene must arrive at this strategy from the point where the gene originates. At the start of life, some genes were simply configured with more stability than others were, and that sets everything else in motion.

It is said that small asymmetries of space-time during the Big Bang, eons later produce a universe of stars, galaxies, and heavy elements. In a similar way, in biological evolution, asymmetries in early genes produce billions of years later, a biota teeming with diversity.

[59] Of course, even a thinner wing or a sharper claw will need other changes at other loci.

## 2.11 Phylogenic Evolution

This chapter, like many, was also written earlier. Before there was an explanation of punctuated rates, there was an attempt to fit evolution into a logical category. For instance, decades ago the evolutionist Theodosius Dobzhansky wrote that "evolution as a whole doubtless had a general direction, from simple to complex ..." This was a general perception, but now much disputed.

This perception is not correct in a sense that everything that begins simple evolves into something complex. Rather, life evolves in layers. A layer of simpler life, for example, of microorganisms, exists first. Of the billions of microorganisms, only a few coalesce into a complex branch of larger cells, which radiate as a new layer. A small number of the larger cells join as organisms, which radiate as a higher layer again. Organisms themselves evolve in complexity over time. Simple plants and animals exist first, and many forms persist unaltered for hundreds of millions of years. Then a narrow lineage of simple organisms evolves into complex ones. If the remaining simpler organisms cannot compete, or are wiped out by extinction, a lineage of complex organisms radiate again. Over the history of life, these cycles appear as a progression.

Is there a name for this evolutionary progression over geologic time from simpler organisms into more complex ones?

Well, there are different names for processes of evolution that are observed on a large scale. A popular name is *macroevolution*, although this describes a scale of events rather than a process or a progression. For example, extinction caused by an asteroid strike is a *macro*-scale event, while a change in moth coloration is a *micro*-scale event. However, this should not mean that a *progression*, such as the evolution of new phyla, results from a *process* of "macroevolution". This is like the debate over the term "punctuated equilibrium". The concept of it drew attention to the unequal rates of evolution observed in the fossil record, which was valid. Yet mere use of the term 'punctuated equilibrium' did not really explain why the rates were different, which was controversial.

There are other terms too that refer to large-scale effects, but also do not explain what causes them. *Orthoselection* means selection in one direction, such as a whale always selected for larger size. *Anagenesis* is another term, which means modification in one line without speciation, whereas in *cladogenesis*, lines speciate via modification. Terms such as *monophyletic* or *polyphyletic* might infer a process, but strictly, they are a method of classification.[60] (Correctly, derived taxonomies should be monophyletic. Labeling a dolphin as a fish is a polyphyletic taxonomy, but

[60] Monophyletic taxonomies are grouped by *homology*, and polyphyletic taxonomies are grouped by *analogy*, but (my view) analogy is an incorrect basis for taxonomy.

as an error.) The term *phyletic* evolution means evolution of new novelties, but with no delineation of why the processes are different; just how they can be categorized. We can describe large-scale change with a term such as 'species evolution' but this too might not logically follow its definition. It is like the phrase; evolution of the species. Different species evolve, but a new species has not evolved until its lineage divides in two, or unless there was once a past species, distinct enough from the present one, but then it is no longer just one species.

In this sense, species also do not do other things, such as *radiate.* We could say that birds evolved in a line; then radiated. Yet we could not say that about a species, because even if a species evolved in a line, it could not radiate and remain one species (with one notable exception).[61] This problem deepens if we formally group species in higher taxonomic units such as families or orders.[62] Such units can evolve and radiate (just as mammals evolved and radiated) and remain one unit (mammals stayed mammals). However, such units are composed of species, which again, cannot evolve or radiate and remain one species. This leads to another problem. Evolution often proceeds in one direction, but then it stops; an effect called *saturation.* However, no taxonomic category of a species, order, or phyla, encounters an effect called saturation.

To explain saturation, and other effects of evolution, I will redefine another evolutionary classification as a *phylogeny.*[63] A phylogeny is not the total organism, or a collection of it. It is the underlying structure composing all living entities.

To explain, a species, family, order, class, or phyla, is an *inclusive* set in logic. (The set *includes* every member). By contrast, a phylogeny is an *exclusive* set of attributes shared among individuals, species, orders, or any category, including all life. The *class* of birds includes the albatross and the blue finch warbler: The phylogeny of birds *excludes* attributes that make a blue finch different from an albatross, although it includes underlying features that birds share, such as feathers. The phylogeny of great apes includes all the shared attributes of great apes, such as hands and stereoscopic vision. By contrast, the human phylogeny is a subset of the phylogeny of great apes, coupled with exclusive human attributes such as upright stance or a large brain.[64]

---

[61] The point is that humans radiate while staying as one species. It might be possible to prove human uniqueness this way via logic. (Also, see bottom footnote.)

[62] For instance, the human taxonomy is Eukarya, Chordata, Mammalia, Primate, Homindae, *Homo*, and *H sapiens*.

[63] Some authorities replace this with phylogenetic, referring to the genetic makeup. I prefer a more traditional use of the term phylogeny.

[64] Note; whereas the human *species* is a subset of the primate *order*, the primate *phylogeny* is a subset of the human *phylogeny*. This is consistent with logic, where the set is always the subset of the union, but the intersection is the subset of the set.

I shall now extend this, to propose that the progression by which any underlying structure evolved is its *phylogenic* evolution. Except that unlike inclusive sets (species, order, or phyla) a phylogeny can radiate, saturate, or encounter boundaries and penetrate them without any of the logical traps of the other categories.

Except that if all of life evolves along the easiest path, this path too will become, we call it, *saturated.* This saturation of the easy paths along which life evolves becomes, I call it a *phylogenic* barrier. When life strikes a *phylogenic* barrier (there are no easy changes left) competition drives organisms to seek a less-easy paths. This can result in changes, causing it to branch in new directions, or to evolve in new layers of complexity, before the cycle repeats again. All change bears a cost, but some changes are less costly to enact. New phylogenies evolve from hard-to-enact changes, but in any epoch, a phylogeny can be exploited.

First, phylogenies must be from the most recent common ancestor (MRCA), via unbroken descent. For example, birds have feathers as a trait, and birds and pterosaurs both lay eggs. To be a phylogeny, though, the MRCA of birds had to have feathers, and to be a shared phylogeny, the MRCA of birds and pterosaurs both had to lay eggs. On the other hand, birds and pterosaurs both have wings, but because the MRCA of birds and pterosaurs did not have wings, they are not shared by unbroken descent. Still, pterosaurs and birds do share limbs, vertebrae, heart, and lungs by unbroken descent. Even so, the MRCA of pterosaurs and birds was a dinosaur, so in some taxonomy birds are dinosaurs. Sharing lungs goes back to the first amphibian, which was the MRCA of all reptiles, birds, and mammals. Similarly, the Diapsida skull is exclusive to birds and dinosaurs, but not to mammals, or other reptiles such as turtles. In this case, the MRCA of the common phylogeny of birds, mammals and turtles goes back at least to the Permian.

If anything, phylogenies exist in space rather than in time.[65] Species live and die, as individuals do, but their core attributes build in layers of complexity, which persist in time. The ancestral species of humans are extinct, but the ancestral *phylogenies* of humans persist from species that preceded humans in time. One (overstated) evolutionary argument is that "given sufficient time" any change is possible. In truth, the amount of change always depends on the available pathways, more than the total available time. Among adaptable phylogenies, new species can evolve in a few hundred generations, which is a short time. Yet the available phylogenies will determine what can evolve and how fast.[66] In "no time at all" limbs can adapt as flippers, wings or legs. Over an unlimited time,

---

[65] Think of a "tree" diagram without a timeline (next page) as representing "space".

[66] Not all phylogenies persist. If there were ever a six-limbed vertebrate, its *phylogeny* is extinct in both space and time. Interestingly, if any phylogeny disappears in time, it will disappear in the space of any "time intersect" that came later.

though, a mammal could never evolve feathers, and a bird could not evolve fur, (disallowing imitations).

The other advantage to phylogenies is that they show the attributes that are truly altering in evolution. If you assumed that all genes alter at average rates, then all attributes alter with equal likelihood. However, a *phylogenic*[67] tree indicates why some changes are just superficial only. Fig 2.11.1 illustrates the phylogenic evolution of a shark, ichthyosaur, and porpoise. In 350 Ma the phylogeny, leading to the modern shark did not alter much, but the phylogeny leading to the porpoise underwent huge changes. Sharks and dolphins share an *analogous* appearance, but one is a fish, which existed for 350 Ma. The other is a mammal, which adapted to sea life within the last 50 Ma. Sharks and porpoises share a phylogeny only of the MRCA of fish and reptiles, 400 Ma ago.

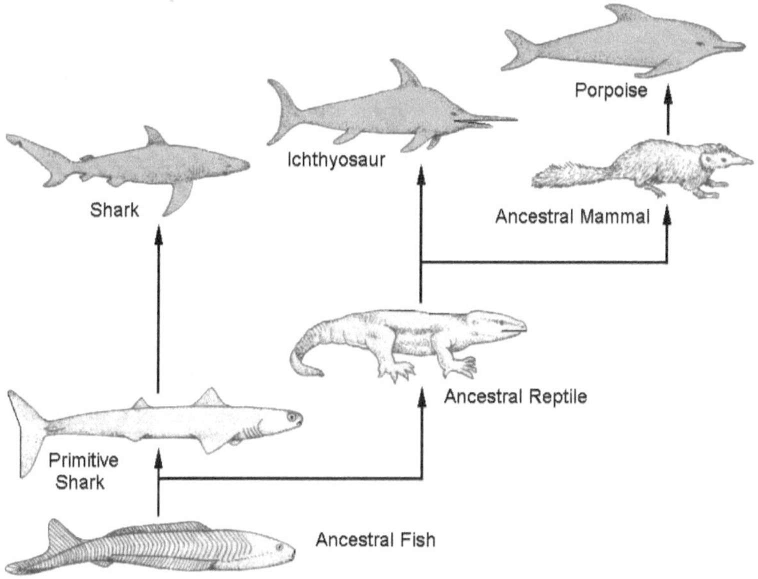


Fig 2.11.1 Phylogenic evolution is of the underlying structures that organisms share. The shared phylogeny of a shark, ichthyosaur, and porpoise is of an ancestral fish. (Redrawn from Strickberger. Similarity of marine shape is 'convergent evolution'. )

[67] The term *phylogenetic* tree refers to genetic branching. Here *phylogenic* tree means a tree of all attributes, including genes, conserved across different groupings.

This diagram illustrates the relationships, but it also indicates the differences in genetic distances from the ancestor. The genetic distance from the MRCA of dinosaurs and mammals over 200 Ma, was less for the dinosaurs for the same period.[68] Notice that total genetic 'distance' from the ancestral fish is short for the shark, but large for the porpoise. Between the shark and porpoise, the lineages diverged $10^8$ generations ago. Genes expressing a *phylogeny* (such as a vertebrate body plan) are conserved genes that mutate slower than this rate ($10^{-8}$). Many genes mutate faster than $10^{-8}$, but these express differences (not homologies). There would be a higher total of faster-altering genes in the porpoise than the shark, because the porpoise diverged further from the original type, and it accumulated newer novelties. [69]

Now consider the reason that phylogenies can *saturate.*

It is tied to how far evolution can proceed in one direction before it reaches a barrier. There are physical limits (such as the effects of gravity) to alteration in length, shape, or size. (It can only be so big, so small, etc.) Yet there is no limit to variation. The length of a worm can reach a physical limit, but it could also evolve short again. Although there are physical limits to how short or long an organism can be there are no limits to how much the property 'length' can vary.[70]

It is the same effect with variations in size, shape, texture, or color. This can come in almost infinite combinations. An animal species might evolve from being brown and small, but then into brown and big. Then it could evolve to be small again, but change in color.[71] There are endless traits at thousands of loci, each capable of variation. In the sense that variation of climate or geology is an endless process, variation to length, shape, color, or behavior of organisms is infinitely variable. This is why the possibilities of change in a species can never saturate.

If the possibility of change in species can never saturate, consider then, the following. A species cannot saturate its possibilities for change, so it seems logical that an order, class, or a phylum, which is a *union* of the species that they contain, could not saturate either. Now we have a new classification, a *phylogeny*, which is the *intersection* of the species associated with it. The question of logic is; despite how a *union* of sets can never saturate, could an *intersection* of sets saturate?

For instance, there can always be a new species of fish, and enhancements to existing fish lines. However, no fish can add a second

---

[68] This is why the ancestral fish was most fit, because it spawned the entire line.

[69] Many ancient creatures accumulate large amounts of junk DNA. Even so, useful genes are of greater variety for a complex, newer evolved creature like a porpoise than a shark.

[70] Nor is variation of a property such as 'length' progress towards completion or perfection. It is just an oscillation around physically constrained node points.

[71] There is much controversy over this. Mendel considered that characteristic genes were not associated. However, characteristics such as eye color and wing length can be grouped in chromosomes, such as in fruit flies, though the linkage can drift apart.

atrium or evolve a more advanced heart. It occurred with lungfish, probably under great pressure, with a type pushed into an early peripheral niche. Yet the two-chamber heart design for fish is now saturated. This is also not just for fish, but each heart type for fish, reptiles, mammals and birds seems to have evolved as far as it can in one direction.

For example, the first fish to add a third chamber to the heart already had primitive lungs. A three-chambered heart only works if there is extra plumbing to make it work, and the plumbing has to evolve to allow it to work the first time. That is why a three-chambered heart in a fish is rare, and cannot compete directly with the two-chambered type in most environments. Three-chambered hearts require a higher metabolic rate, at more energy to sustain, and this is not always viable in the ocean where life is easily sustained at lower metabolic rates. Only after it migrates onto land, away from fish, can the three-chambered heart phylogeny *radiate*. Yet once it has radiated on the shoreline, or land, and dominated that range, this makes it hard for a rival to repeat the cycle again.

Moreover, having evolved one, two, three, and four chambers, heart evolution has now reached an upper limit of practical design. It is similar to petrol engines. There are two-stroke and four-stroke petrol engines, for broadly the same reasons that there are two and four chamber hearts.[72] Nevertheless, while there can be many styles, sizes, and ratings of petrol engines, there will never be a six or eight stroke engine of any practical use. Similarly, while there could be freaks or mutants, there could never be a viable five or six chamber heart for present life on Earth. In birds and mammals, from separate evolutionary paths, circulatory efficiency is now at an optimum of basic design.

While any attribute can alter, phylogenic evolution is irreversible. A four-limb body plan will never de-evolve to two or no limbs, although limbs can atrophy, like on a snake. A two-chambered heart can evolve into three-chambers, but it can never *de-evolve* back two chambers. Reptiles or mammals who have readapted to the sea, never readapt the two-chamber heart. Genes expressing the material "feathers" will exist in the genome of bird descendants. Featherless birds can exist as an adaptation, but any featherless bird will retain the genetic code to re-adapt feathers again.

This principle that competition could inhibit certain attributes from evolving was mentioned long ago by John Maynard Smith:

> "The absence of any new body plans need not indicate a loss of evolutionary potential. It is more plausible that, once a great variety of complex animals existed, further attempts to evolve a complex body plan from scratch were inhibited by competition. In the same way, we think that the origin of life

[72] In both cases, the problem is preventing the oxygen-depleted line from contaminating the fresh, oxygen-rich supply. Creationists though, will point out that the two-stroke did not evolve into a four-stroke, like chambers for the heart, but was a later simplification.

> itself was a unique event, because once living organisms existed they would rapidly destroy any later beginnings."

This paragraph elucidates that evolution can be inhibited by competition. Evolution of complex new types is never impossible because it happens; but it is statistically less likely than adaptation within a type. Less likely events in evolution do occur, but they take longer, with less chance to be viable beyond the first time that they evolve.[73]

In summary, there can be a one, two, three, or four-chambered heart, but not a five-chambered heart for life on Earth. RNA evolved into DNA, then saturated. RNA first coded for a few amino acids, then 16, then 20, but the progression has now saturated at 20.[74] This is because change not only comes at a cost, but it evolves means to adapt at lesser cost. For instance, there is no need to re-evolve the genetic code each time that an organism adapts, because organisms with the modern code already adapt faster than a new code could evolve. In the modern era, no reptile could re-evolve endothermic regulation in competition with mammals, because that novelty already allows mammals to adapt at a much lesser cost than to re-evolve the novelty again.

Unlike for species, phylogenies *saturate*, causing large changes in evolution. Organisms best adapted to their suit of traits, the fittest of existing types, maintain existing ranges. Yet newer organisms will be pushed from existing ranges, where to survive they must evolve new phylogenies, regardless of the costs. If the planet can sustain the cost of continued evolution, it continues until every niche on the planet able to support life is penetrated, and every phylogeny is saturated.

When this occurs, we need a radical new form of phylogeny, if life is to evolve further, in an unprecedented way.

---

[73] The barrier against radical change is also cost, and viability of less thermodynamically probable changes against simpler changes (Chapters 2.1 and 2.4).

[74] Apparently, there is one more, extremely rare amino acid, pyrollysine, which evolved for a specialized need in a few microbes.

# 3.0 Human Evolution

## *3.1 Why Humans Evolved*

In this section, I will explain human evolution.

Now, there are many ideas why humans evolved, but let us divide this into facts and theory. Assume for now, that humans evolved as per the scientific fossil record. This is often being updated, but if you must check facts against the latest fossils, just do it.

Instead, let me give two new theories, why humans evolved.

The first one I call *Maximizing Options*. Now, I do not have the math, but this is an old theory, by me, of how humans had evolved to maximize the options of behavior. The other theory, also by me, is that the human brain developed to include both the evolutionary impulses, in how we behave, but also to include logic, as a method to behave. There is little source on this, but I discuss it next chapter.

Now my first book was called the Theory of Options, and it was well received by a few people, but it never gained traction. I repeat the theory here. It was that it had proven difficult to produce a theory, which could combine both human evolution, and psychologically motivated behavior. This could be met by a theory where humans evolved to maximize the options of behavior. This allowed a new theory of psychology, because in everyday behavior, humans often act in ways that maximize their choices. Human behavior makes sense by individuals trying to maximize choices, rather than maximize the passing of DNA. On the other hand, human evolution could be explained if we allowed that it was also along a pathway that maximized the options of behavior, for a given cost to evolve, for various cases.

For example, if we wonder why humans evolved a hand delicate enough to perform brain-surgery, it was not for that end. So, the question is turned into; which design of a hand would maximize the options of behavior? Looking into this, we discover that the modern hand offers the most flexibility for a given cost to evolve from what it was (an ape hand) into what it is. A hand more delicate than a human one is conceivable, but all change bears a cost. A hand that evolved maximum flexibility for the least evolutionary cost of change happens to be capable of performing brain surgery, although it might not be delicate enough for other tasks. It is similar with body hair. For maximizing options, a body devoid of hair offers an extra option of going naked, if other means can be found of keeping warm. Yet the body need not be fully naked. It must balance flexibility against the cost of evolving. This leaves a human body mostly devoid of body hair, but with some residual covering.

In these terms, "options" is a word with no quantifiable meaning. It is not like maximizing DNA, which is factor that can be measured. Rather, to quantify the meaning of maximizing options is more difficult. Broadly, when humans evolved, life was already *saturated* with other species that could rapidly adapt. Fit humans had to *maximize* passing on DNA, but for *minimum* changes in genes against a *cost* to evolve. What say, is a claim that life was already *saturated* with adaptable species? Why would this force one species to evolve along a pathway to minimize the changes in genes as a way to maximize passing on of DNA? Why would not an organism attempt to maximize passing on of DNA alone?

Again, there are many theories why humans evolved. Perhaps, after reading it, you still believe in God. Yet there are facts about how humans evolved that must be correct. The next diagram is a recent picture, from Wikipedia, of one scenario. Again, this is the fact of how life evolved. It is checked by fossils, dug in many places, put in a rough order. Whatever theory you propose from this, to explain the fact, it must conform to the fossil record as recorded. Either this or your theory does not work.

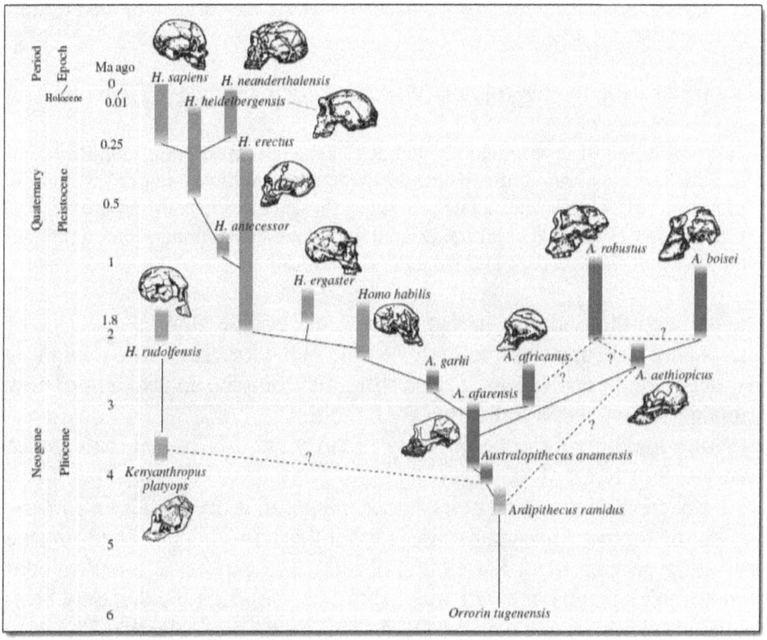


Fig 3.1.1 Human evolution involved many branches over 5 Ma. See also Fig 1.1.6, how different species evolved in parallel.

Notice, there are three major branches in human evolution, between a lead to *Homo sapiens*, and *Rudolfensis* and *Robustus*. There are minor branches, between *Homo sapiens* and say, *Neanderthals*. The splits were for many reasons, but you must explain these. Again, if your theory cannot explain the facts, there is no point.

For instance, if you claimed that a shark evolved along a pathway that maximized adaptation for a large marine predator, nobody doubts that this pathway existed. If you also claimed that an albatross evolved to maximize fitness for deep ocean flight, nobody doubts that such a niche is available. Even if you claimed that humans evolved along a fitness pathway that optimized bipedal life on the plains of Africa, there is an explanation for this. Forests were shrinking, and there are many reasons why a former forest-dwelling ape would do better on the plains with an upright stance.[75] Because the climate was fluctuating, there are also reasons why flexible behavior would be an advantage. The renowned Jacob Bronowski had once explained:

> "Human evolution began when the African climate changed to drought: the lakes shrank, and the forests thinned out to savanna. And evidently it was very fortunate for the forerunner of man that he was not well adapted to these conditions. For the environment extracts a price for the survival of the fittest; it captures them."

In a similar vein, the neurologist William Calvin argued:

> "The evolution of anatomical adaptations in the hominids could not have kept pace with these abrupt climate changes, which would have occurred within the lifetimes of single individuals. Still, these incremental environmental fluctuations could have promoted the incremental accumulation of mental abilities that conferred greater behavioral flexibility."

The point to these claims is that humans did evolve along some type of fitness pathway, one that required flexible behavior. The issue though, is that interesting as this claim is, this still might only be an accident of how primate life evolved in those varying conditions.

Now, maximizing options is my broad view, but human emergence involved four types of selection.

Classic Darwinian selection is *environmental*. It occurs because in any population some individuals are randomly born better fitted to the prevailing environment. For example, light skin can better adapt to cold climates. Organisms not well adapted fail to reproduce, so that only best-adapted individuals can pass on DNA. This method of selection is clearly effective, except that it is somewhat slow, in that fitness emerges as a statistical trend from many random failures. This trial-and-error fitness also does not allow behavior to affect selection.

The next type of selection can be called *behavior first*. It means that animals often adapt behavior faster than they adapt biology. In this way, early giraffes did stretch their necks to reach leaves. Just that following

---

[75] One researcher has claimed that pre-humans first learned to walk in trees anyway.

behavioral adaptation, genetic selection will better adapt a species to its new behavior. Giraffes with long necks were selected first behaviorally by individuals seeking the leaves on tall trees. Darwinian selection was a genetic follow-up of behavior "first".[76]

The third type of selection is *sexual*, where partners select from consent or preference, rather than physical dominance. This increases selectively, because it can bring many behaviors, effectively of the whole group, to focus on selection. This way, not just the partners select, but as in human society, group preferences influence individual selection. This causes evolution to speed up, by making selection less by trial-and-error, and more by using the lessons of group experience. All large animals, even fish and reptiles, use some sexual selection, but in humans, it seems to have become the pre-eminent form of selection.

The final type of selection is splitting-up.

Most evolution is by *cladogenesis* in which species divide into sub-groups. Splitting-up speeds evolution, by bringing behavioral pressures on the selection process. There was much migration, and sub-branching into differentiated groups during human evolution. All species divide into groups, and as groups isolate, they speciate. One basis for humans groups was behavioral, when adventurous migrated to search for new food sources, while the cautious stayed with the existing food supply.

Splitting-up could also explain a propensity to maximize options.

Suppose that each time a human sub-species evolved, individuals differentiated into two mixes of genes; "specialization" and "options". Each mix gambled on how the environment would alter. "Specialization" gambled that the environment would alter slowly over millions of years, allowing time to specialize to any change. Yet this is not what happened. From four to three Ma, an environment might have been relatively stable, but then it began to fluctuate. Temperatures rose and fell, rainfall varied, and forests shrank in periods as short as centuries. At each change, the lineage adapting by 'specialization' was eliminated by failure to compete with individuals that possessed more versatile 'options'. Even after genes for specialization were eliminated, though, genes that caused groups to divide passed on in both groups, so the process could always be repeated. Maybe 'strong but dumb' individuals stayed in one place, while 'slender but smart' ones split away. Moreover, during an era of climate change, genes had a way to guarantee survival, by selecting individuals to adapt to survival in any environmental condition.

In hominid evolution, versatility triumphed over specialization many times. It was mentioned how a skin devoid of body hair offers the most options among skin coverings, but another example is diet. The *robustus* lineage was specialized with heavy jaws for chewing roots and nuts. This

---

[76] Genes "first" is that a random mutation allowed individuals to adapt new behavior.

was too specialized, because if conditions changed and a different food source was needed, versatile hands and a learning brain could best adapt to these. (If the only source is coarse food, versatile hands and a clever brain can find ways to soften it.) Similarly, to how Neanderthals were wiped out because of specializing to a climate, the *H. robustus* species was eliminated because it became specialized to a niche diet. There were many branches in hominid evolution. However, the creature that kept its evolutionary options open was the one that was able to proliferate. Maximizing options, in turn helps explain why today there is only one intelligent species. [77]

When other species divide, one result is each group finding a niche to which it can specialize. For adaptation into a specific niche, there will be as many varieties as there are environmental niches to accommodate them. Selection will also be forever variable to the environment, so that adaptation will never saturate. It is an open process. However, human ancestors were not specializing to one niche, but were generalizing to all environments. One species can always specialize to a desert, say, better than it could to a forest. However, to generalize across all environments, especially fluctuating ones, one group alone will be better adapted to that need than the rest. This way human evolution became a closed process. It continued until one species became better adapted for competition and survival than any intermediate type, across Earth.

One early split was evolution of *Ramidus*, five million years ago, and then followed by *Australopithecus*. The achievement of early species was reorientation of the pelvic structure for standing erect. It might have been survival of the fittest, for 'better walkers' to split from slower moving cousins. *Ramidus*, and later *Australopithecus*, did not developed tool making, or cultural adaptations such as an increased brain size. Although divergence occurred in early hominids, direct fitness was a strong force. Sexual selection occurs in all higher animals, but in hominids, it would be reinforced by sexual aesthetics. This included a mechanism to select away from the body of an ape.

About 2.5 Ma ago, tool making emerged with *Homo habilis* as an organized activity. By 1.5 Ma ago, *Homo erectus* began migrating from Africa as far as Asia. Tool making and migration require cooperation, so environmental selection between individuals must have been giving way to competition between groups. Even so, group and sexual selection only operate once environmental selection has provided individual variation. Change such as increased cranial capacity occurred via groups splitting-up, which resulted in migration and extinction of rival species. Adaptation to a local environment (racial adaptation) is likely from "behavior first"

---

[77] Neanderthal man now seems to have been an Ice Age adaptation, with a sturdier frame better adapted to cold, but was out-competed by the more versatile humans.

selection. Other modification would be via environmental selection. Sexual selection occurred all the time. [78]

In human evolution, splits occurred in the existing population, and their endless branches. Whichever hominid line succeeded, the genes that forced splitting, in all branches, would survive. After five million years, an optimal line emerged from the branching that could spread almost unaltered in the huge numbers that we see today. Essentially, if a population remains small in a fertile world, it is a specialist rather than generalist adapted, so it must keep adapting or risk extinction. If a specialist population spreads, as the migrating groups encounter varied environments they will speciate anyway. If one line becomes generalist though, optimally adapted to all environments, the line will spread widely, supplanting niche adaptations. If a species is widely spread today as a unified species, like humans, it does not mean that this is not the result of numerous speciation events in the past. In whatever case, the gene is always better off if it forces speciation on opportunity, so the logic for a gene increasing its spread supra the 100% of a single population by speciation still stands.

What is to be made of the point of emergence occurring near the end of the Ice Ages? If fluctuating climate was driving human evolution, when the climate stabilized (we are not sure that it has) had humans emerged optimally adapted? This question is important if one is claiming that the evolution of intelligence was a general effect. The cold might have accelerated human evolution, but maximizing options was a process that continued regardless of climate. The struggle was between emerging new species and sub-species not yet optimally adapted. This reached a culmination, because no sub-species remain. Even if the African boiling-pot had shut prematurely, if there were any potential left to human evolution another group might have thrown it up.

Emergence was complete when forces of evolution in African became counterbalanced, once the earlier migrated species could resist pressures from African migration. By then human evolution into a new species was complete. An end condition of not slow biological evolution, but fast culture evolution would take over. Significantly, when cultural evolution did take off, it was among migrated species in Europe and Asia. This reinforces the premise that stay-at-homes were the more successful Darwinian types, but those forced to wander were the more versatile adaptations. [79] This is controversial though, because the distribution of culture with migration is not a uniform effect.

---

[78] I made much of this before. Sexual selection was important in human evolution, a point noted by Darwin. It is just that this topic has been discussed so I will not repeat it here.

[79] Eskimos or aboriginal people migrated far, but did not evolve significant culture outside of nomad existence. South American Indians migrated further than North American ones, but evolved a more intricate culture, so there must be several factors at work here.

Still, migrating species were becoming environmentally specialized at their destinations, while not evolving much beyond attributes brought from Africa. The final split from *Homo erectus* into *Homo sapiens* less than 500,000 years ago, involved one final migration replacing all other species. Over a five Ma period, many biological adaptations were tried, but for reasons mentioned, only one final variety emerged. When cultural evolution takes off, the species continues to evolve biologically, but not into a new species. For humans, the point of emergence may have been as little as 50,000 years ago, near the end of the last Ice Age. Humans had by then maximized their biological options. Beyond that no higher species had emerged from the boiling-pot of Africa, to displace species already migrated. From that point on, a uniformly human species began slowly to occupy every continent on Earth.[80]

Especially, as life evolves phylogenies become more versatile. For example, mammals are already versatile, but primates evolved higher in versatility beyond that; litter sizes went down, paws evolved into hands, backbones and limbs became more flexible, brains became larger, and behavior became social. This allowed for more rapid adaptation. A hand is more versatile than a paw (it can grasp tools); learned behavior can adapt faster than inherited behavior (genes do not have to alter), groups can learn faster than individuals. Moreover, a brain that can learn not only learns; it can evolve faster for a lesser genetic cost of designing each neural circuit for reflex. Once the ape lineage could adapt faster than other species; phylogenies of species that could not adapt as fast became saturated against comparable costs to evolve.

In this sense, when a trait in one phylogeny (paws in cats) cannot evolve at the speed or minimal fitness cost of the trait in a rival species (hands in apes), that phylogeny (cats) is in saturation. When phylogenies saturate, they then divide into sub-phylogenies, which also saturate. Once great apes had hands, a large brain, versatile limbs, stereoscopic vision, varied diet, and crude tools, few other novelties could be squeezed from the evolution of large animals. That too is why great ape evolution resulted in a general saturation for large animal life on Earth.[81] This means that once all major classes, orders, families and genera of large animals were established, nothing could evolve a significant new novelty, faster than a great ape could evolve into a human.[82]

---

[80] I now consider that racial specialization beyond Africa was more complex than this. Many cultural factors effected different types of adaptations.

[81] Great ape is a family, not a species. The human species did not evolve from a modern ape but from a common human-chimp ancestor. There is a great ape *phylogeny* shared by gorillas, orangutans, chimps and humans. There are also independent *phylogenies* of humans, chimps, and so on. All the phylogenies are very close to general *saturation*.

[82] Unfortunately, this is now bringing tragedy. Humans have already destroyed many species such as the Mammoth, which could not adapt fast enough, and threaten many more.

On the other hand, the preferred model for explaining adaptation is where a favored allele sweeps a change at a single locus. The limitation with this model is that once the new allele has swept to fixation, the population stays the same species (all individuals contain the same new allele). However, for hominid evolution evidence is overwhelming that there were at least several large splits.

The anathema with group selection is not that it occurred, but lack of a selfish-gene model (the gene would not act for "the good of the group" but see note[83]). The gene always acts for itself, but the standard model has a rule that maximum distribution is 100% in one population. In my model, a gene can distribute beyond the 100%, if the population divides. In human evolution, populations split many times, and hominid groups were selected as whole populations. Population-level selection is proven by the fact that only *Homo sapiens* persisted, and all the other groups were lost to extinction.

Again, a theory of humans maximizing options is not mathematical, but there is a model in control theory, where criteria are set so that the result is towards some maximum. I believe a related problem applies to human evolution, but I do not have the genetics or the mathematical skills to develop it. On the other hand, I do not think one can claim that human evolution was merely the result of genetics and might have gone a different way. Something made the human evolution unique, in the world. Maximizing options is one theory that needs further development, for its effects to be fully realized.

Next, consider the next development. How one part of maximizing options, humans acquired a unique brain.

---

[83] This is another statement made with the knowledge at the time. Research has since uncovered that actually, where animals survive best in groups, there is selection towards helping the group. Suppose that a troop needs a minimum number of males to hunt and defend for the troop. Then if a dominant male killed all its rivals to get all the females, its genes would still not survive if the troop were wiped out.

## 3.2 Logic in the Human Brain

Why can the human brain be logical?

Is it philosophical? Did the brain, designed by evolution, now modify by inputs that are outside of biology. Is it, however, further down? Can people who do not philosophize, still do math? Not everyone can, but there is an input from math that shapes how the brain is made. This has always bothered me. Ancient writers noticed it. The Muslims observe it (though I cannot agree with their conclusions). There has always been some conflict between truths derived by logic, and more ordinary truths, such as a belief in God. I do not myself see an impeccable logic can produce an argument for God, though this is my opinion.

Today as well, the criteria for judging is strict. A person, for example, has an argument that might reveal God. The issue is that all such reasoning will go via a person who does not believe in God, and criteria are strict. It is useless to philosophize that all people see this, or that many people do, or some other factor plays a role. Unless the reasoning passes a criterion, agreed by many people, it leads to further support for the case. Therefore, one can claim that proofs of God, or some other effect cannot meet criteria. You can lower the task for a mathematical proof, but all the bets are now off. Many conjectures are possible, but only certain criteria pass that meets an agreed standard, will pass the test.

If we move the argument from mathematics to science, here are more possibilities. Not everyone can agree on all the facts. Certain criteria from philosophy say, to science, are not as agreed. We can see this in Evolution. It is not logical that certain criteria applied to how the universe was formed, but those same criteria did not apply in evolution. To form the universe, certain criteria were moved forward. Residual properties formed a case of an 'imaginary' effect carrying into the evolution of life. If we consider evolution itself on the other hand, this extra effect is not included. There is proof of it in molecular selection, and proof that molecular selection in how genes form. This is in the records. But if we turn to evolution the extra activity is not there. We can see it in the records that gene selection had played and early role. However, if we shift to general evolution, also in the records, molecular selection has been overlooked. An incompatibility is now here, that molecular selection must exist first, as a precedent about how cellular selection unfolded.

Once we turn to humans, we must explain why the logic arose. The human brain is logical, so how did it get there? The issues us that hominids could have done everything needed, hunt, survive, procreate, migrate and multiply with a smaller brain. Yet the brain, with 2.5% of human body mass, consumes 20% of the body's energy. Rival hominids with smaller brains had more energy for hunting and producing offspring, and those offspring were easier to bear because the head was smaller. However, to survive, be fit, and pass on DNA on the plains of Africa, reproductive fitness was not just measured in total offspring or catching prey. The brain needed to be capable of logic, as mathematics, music, and philosophy. These only developed as humans evolved. Especially, it was only once a capacity to manipulate mathematics evolved, did the capability for truly logical thinking also evolve.

There are also many theories, but they compare attributes from other species where they might not apply. For example, there could be a simple correlation between brainpower and human reproductive fitness, because intelligent parents can better provide for a long-term well-being of their children. Stag elks or peacocks cannot plan the well-being of offspring, so for these a mating display could be a better guide, even if it overuses mating signals. Further, mating displays are more pronounced in species with harem-style mating, where one favored male command the females. It is contentious, but there is little evidence that tribal life for humans used harems. Providing that a male hominid is intelligent, he usually gets to mate. In human society, a male with the largest head does not commandeer all the females, the way a stag elk with the largest antlers,[84] or the peacock with the most colorful tail might.

Further, when explaining the large brain, defenders are trying to confute skeptics, to prove that a large brain can be explained by gains for single genes or alleles. If the human brain is larger than a chimp's, then it must be filled with additional neural circuits. If humans exhibit complex behaviors, the extra circuits will explain these. Therefore, all you need to do is to identify a slight advantage for each of those extra circuits, and the problem is solved. The intrigue of this approach is that whatever hominids were doing during the period when the brain was evolving, it was not mathematics or philosophy. Hominids were hunting, fighting, and producing offspring, so the evolution of extra neural circuits needs to be explained in those terms.

To suppose that humans evolved along a pathway that maximized the options of behavior, it sounds like a generalized outcome. We suppose that on a planet that can sustain prolific life, *saturation* will set in. When this happens, one line might evolve in a way that maximizes the options of behavior, which on Earth led to intelligent life. This infers that from a prolific biology, the evolution of intelligent life naturally follows, but this too is disputed. As Steven Gould explained;

> "At any of the hundred thousand steps in the particular sequence that actually led to modern humans, a tiny and perfectly plausible variation would have produced a different outcome, making history cascade down a pathway that could never have led to Homo sapiens, or to any self-conscious creature."

---

[84] Some researchers think it is the Stag's bellow rather than its antlers that attracts females.

It returns to explaining the humans were a mere by product, but again I ask the point. I incline towards a view that from prolific life, and the right planetary conditions, intelligent life has a chance to evolve. Nevertheless, other indications are that it is difficult to get a favorable combination of planetary factors that would allow long term, stable development of life, until we find out more.

The favored line of reasoning is to liken the brain to an overgrown organ, such as a peacock's tail or a stag's antlers. Such organs evolved to their large sizes in a runaway competition, between rivals for partners, or for dominance in the herd. Using this model, Richard Dawkins wrote;

> "It is even possible that that man's swollen brain, and his predisposition to reason mathematically, evolved as a mechanism of ever more devious cheating, and ever more penetrating detection of cheating."

A similarly popular view, by Geoffrey Miller is;

> "The neo-cortex is largely a courtship device to attract and retain sexual mates: its specific evolutionary function is to stimulate and entertain other people..."

The game is a mental attribute that will somehow translate into more offspring, then the rival must do better, and so the brain size increases.

A typical explanation is as follows.

The human brain is large, but the peacock's tail is also large, and to attract females. Therefore, the reasoning goes, the human brain is also large to attract females. Yet why would a large brain attract females? The answer is that women think an intelligent father will better provide for offspring, so they need a test to tell if a male is truly intelligent or is just faking. An allele evolves in females for being attracted by witty and entertaining stories, and an allele evolves in males for telling such stories. Any male who can tell witty stories will then get more females, hence more babies, and the allele spreads. If rival males are to compete, they must tell even better stories, so a new allele spreads. As each new allele spreads, the brain grows larger, like the peacock's tail.[85] Steven Gould called them "just so" stories. We are genuinely puzzled over how in the wild a brain could evolve that could invent calculus, paint the *Mona Lisa*, or write the *Critique of Pure Reason*.

Note too that complexity evolves at a cost, paid by the time, energy, and ecological resources of a planet. If life on any planet cannot afford the cost to evolve beyond simple forms, the evolution of more complex life forms will arrest at that point. Earth though, seems to possess a huge capacity. For billions of years life on Earth has not encountered a cost to

---

[85] For this explanation of why the brain grew larger from telling stories, I have cynically noted that Pinocchio's nose grew larger from this effect as well.

evolve further complexity that the planet could not support. We want to know what happens when the resources of any planet can furnish the cost of any level of biological complexity, and why this led to humans as an outcome of prolific life in this case.

The theory now is that the fitness leading to human evolution resulted from *saturation*. Chapter 2.4 discussed a model of saturation called *phylogenic* evolution, in which species evolve around core designs (a four-chamber heart is a phylogeny). Species are continuously variable, so they never saturate the possibilities for further adaptation. By contrast, phylogenies saturate how far they can evolve for a given cost (a six-chamber heart is too costly to evolve). Complexity makes types variable, which allows them to adapt faster for equivalent cost. However, when all types within a range (for instance, chimpanzees in Africa) can adapt a minimal cost of change, that phylogeny *saturates*. This forced human evolution into radically new modes of adaptation.

Even so, individuals have to be fit, and pass on more DNA than rivals. At the start of hominoid evolution, selection was Darwinian, but today it is largely cultural. Yet during human emergence, biological and cultural evolution interacted. It is not fit for humans to evolve delicate teeth and jaws, if they do not know how to tenderize food. It is not fit to shed natural body hair (as temperatures fall) unless there are other ways of keeping warm. There would be no fitness advantage to have sensitive hands, if these have not learned how to fashion tools or grasp weapons. Many examples of this type can be found.

Moreover, while culture arises via cooperation, populations modify via competition. It means that individuals would be forced to cooperate culturally, while also competing for increased fitness. For instance, books often claim that activities such as tool-making increase brain size, yet tool making is acquired, while brain size is inherited. Acquired changes cannot produce inherited ones (or not directly). There must have been other selection processes at work, or humans would not have biologically modified to the extent that they did.

Perhaps the model in this book can provide additional insight. The expansion of brain size, intelligence, and flexible behavior is explained not by newly adapted alleles, but an impetus to maximize distribution of all genes in the hominid genome, at a minimum cost of alteration to conserved gene sequences. True, humans have more neural circuits than chimps: some 90 billion neurons in the human brain, where a chimp has 30 billion. The paradox though, is that human and chimp brains, and the brains of all higher mammals, are expressed by the same number of genes (20,000 genes in 30,000 gene genomes). [86] The reason that humans express the

[86] I am allowing 10 billion 'fixed' neural circuits in both chimp and human, but four times the volume per body weight of learning neurology for the human, and twice the volume of

extra neurons for the same number of genes is that the extra circuits are wired by the learning experience.

A human child has only 25% of adult brain size at birth, compared to 65% for a chimp brain. Most neurons in the human cortex are unwired at birth, but in the first few years after birth, they connect at a phenomenal rate, over 10,000 connections per second![87] Research has also shown that children raised in non-stimulating environments developed brains 20% to 30% smaller than normal size. Even rats raised in stimulating environments develop 25% more synapses per neuron than rats raised in drab surroundings. So, although humans were not learning mathematics while they were struggling for survival, it is also true that about 80% of whatever humans do behaviorally they learn post-natal. The question is not why the brain expanded for the activities that it was learning then (such as how to hunt), but why the human brain needed such a high ratio of learning neurology.

In truth, humans needed a large brain for many reasons.

The human brain can be used for courtship, cheating, to ensure the survival of offspring, and in many ways. Big brains are always better, except that every attribute evolves at a cost. There is an energy cost to maintain any organ, but there is also a genetic cost for it to evolve. If the only puzzle is the cost to maintain an organ, then if this seems high, it is difficult to explain why that organ evolved at all. It is like the peacock's tail. The tail seems a costly attribute to maintain, and migratory or other birds that need high flight efficiency do not have elaborate tails. Yet for the peacock's niche, and its courtship behavior, the genetic cost of an elaborate tail is likely to be low, because it is usually easy to alter the size, shape, or color, of an attribute, by altering a few genes. It is similar to the way a stag elk can grow bulbous antlers, although this requires other changes, like a stronger neck or bigger heart. Such grouped changes though, can also be easy to enact genetically.[88]

By contrast, the human brain is multifaceted. Geometry for a larger head might take small genetic changes, but changes such as the complex birth process, years of nurturing offspring, or changes for increased blood flow to the brain required changes in many genes, and apparently to the human chromosomes.[89] Because change incurs costs, hominids would not have evolved big brains if they were not also forced to keep adapting. (If the forests had not shrunk, hominids might have remained similar to

---

in the chimp. This gives 80 billion learning neurons in the human, and 20 billion in the chimp. Estimates based on brain volumes and might need adjusting for neural density.

[87] I first read this as 30,000 connections per second, but even that might be low. A year is $3.1 \times 10^7$ seconds, and for $10^{13}$ connections, this would in fact compute at the incredible rate of $3 \times 10^5$ $sec^{-1}$ connections for one year.

[88] I wrote this, years ago. Since then Sean Carroll and others have found genetic 'switches' that truly bring about these changes at low molecular cost. (See *Science Daily* 8/22/08).

[89] This is the fusion of chromosomes 2 and 3 and several inversions, mentioned earlier.

chimps.) It is not easy to evolve a large brain, so the question is that if a brain is forced to evolve quickly, which type of brain can expand the most for a minimal genetic cost of change?

My hypothesis is that neural circuits used for leaning offer the most advantages for rapid brain expansion.

When brains first evolved, each neural circuit had to be designed by selection, a tedious process that perhaps required a new gene for each new circuit. There are advantages in 'fixed' circuits, because they are fast and reliable (like fixed logic in computers), and all brains use core logic for crucial functions. The drawback with fixed neural logic though, is that it will quickly consume the available number genes to encode it, if each circuit needs one gene. The evolutionary solution was 'learning' circuits. These can be of a common genetic design, but they can be wired after birth by experience.[90]

Neurology for learning probably evolved in the Devonian, when vertebrates began to venture onto land. The vertebrate forebrain has the most learning circuits, so the primal forebrain was used for smell, as an early skill. When amphibians first ventured onto land, though, the newest skill was walking. This was first controlled in the learning cortex, while earlier evolved skills, such as for smell, were transferred elsewhere. Once walking was perfected, its control was transferred to the hindbrain, to leave the learning-capable forebrain free for newer skills. As vertebrae life evolved all segments of the brain grew, but the forebrain grew the fastest. In early vertebrates the forebrain, midbrain, and hindbrain, each take about a third of the cranial capacity. However, in humans, while all segments of the brain grow, the forebrain grows seven to eight times larger than the other segments combined.

If hardwired neural circuits are reflex, and the soft-wired ones are learned, then each species has an optimal ratio of learning to reflex. I call it a *learning ratio*, of instinctual to learned behavior (Table 3.2.1). Note that humans have 3.4 times the brain volume of a chimp (about 1350cc in man, to 400 cc in a chimp) but four times the area of higher cortex. Then assume that both species possess similar reflex in proportion to body weight, which is about 2.3 cc of reflex brain for each kg of body weight. This gives a three-to-one ratio of learned neurology to reflex for a chimp, but an eight-to-one ratio for humans.

There is also a larger frontal cortex in humans, which occupies 29% of the human brain, but 17% of a chimp's brain. The cortex represents conversion of functions that were earlier performed by reflex. However, the frontal cortex is a further conversion of functions of the middle cortex. This gives a ratio of frontal cortex to reflex of 50% in chimps, but 230%

---

[90] This is oversimplified. All neurons have a slight amount of learning by synaptic facilitation, and even learning circuits can be of 'wire-once' or 'wire-many-times' types.

in humans. This way the human brain could possess almost a 99% commonality of circuit design with a chimpanzee, yet it can still be a radically different brain because of the different ratios of reflexive neurology, to higher and prefrontal cortex neurology.

| **Attribute** | **Chimp** | **Pre-man** | **Human** |
|---|---|---|---|
| Brain cc | 400 | 900 | 1350 |
| Body kg | 45 | 54 | 65 |
| Body gram/Brain cc | 113 | 60 | 48 |
| Learning cc | 297 | 776 | 1201 |
| Reflex cc | 104 | 124 | 150 |
| **Learning Ratio** | **2.9** | **6.2** | **8.0** |
| % Frontal Cortex | 17% | 22% | 29% |
| Frontal cc | 50 | 171 | 348 |
| % Front Learning Ratio | 50% | 140% | 230% |

Table 3.2.1 Ratio of learning to reflex cc is the estimated **Learning Ratio**.

When humans evolved, they needed to learn a huge range of new skills quickly. Walking was still in the hindbrain, but now supplemented by a vast expansion of learning circuits. (So while crawling is still mostly reflexive, for humans walking upright is a learned skill.) Many other human skills, such as speech or tactile dexterity, are also in the learning cortex. Even vision is partly learned,[91] and it makes sense. To perfect the new skills that humans had to learn by selecting a new gene for each skill would have taken thousands of genes and a huge evolutionary effort. However, once the design of a learning circuit is perfected, these simply need be multiplied millions of times for the effect.

In this, human brain has three times as many neurons as a chimp, or seven times more than a mouse, for a like number of genes in each species. Fig 3.2.1 illustrates how it works. Note from Fig 3.2.1, that the human brain case expanded in volume by a genetic instruction. However, the increased volume was not filled with individually designed and selected neural circuits.[92] It was filled with learning neurology, for an increased brain size and capability, but a small genetic cost of change. Fig 3.2.1 indicates relative expansion of the generalized cortex against function-specific parts of the human brain, compared to a chimp. The expansive

---

[91] Tragic evidence for this was discovered when a child lost the sight of one eye when that was bandaged for a minor ailment for the crucial few weeks when the eye "learns" to see.

[92] This diagram shows differences between functionally specific and generalized areas of the higher cortex. (Seeing as the brain stem is reflexive, I have colored this gray at my own discretion, but it was white in the original diagram.)

non-shaded area does not mean that the human brain is *tabula rasa*; a so-called 'white paper' as philosophers once thought. If humans do not study mathematics or play chess, it does not mean that their brain is unused. Research has shown that the brain is used, with most of it for functions, which include activities such as walking, speech, or recognizing faces.[93]

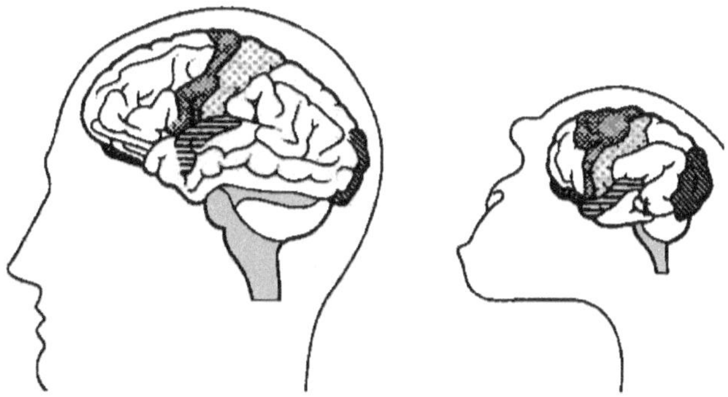

Fig 3.2.1 A comparison of human to chimp brains; The fixed neurology of any brain is roughly in proportion to body size. Yet learning neurology expands in proportion to intelligence. Shaded areas show the learned neurology dedicated to specific functions, and the white area is generalized responses. The general area is 3:1 to fixed response in chimps but about 8:1 in humans.

Yet why did the human brain still evolve so large, to an extent that it could grasp mathematics or logic?

Brains had been growing larger for half a billion years, when human evolution began. The reason was that brains made behavior flexible, so that organisms could adapt at little cost of total genetic change. Learning neurology furthered this trend, by allowing greater adaptation for fewer alterations to core sequences. Once mammals evolved primate brains grew even larger, and hominid evolution triggered a competition among individuals for increased brain size as a means to adapt.

Again, the reasons for competing for increased intelligence are well debated, and everyone has a theory. Courtship, cheating, telling stories, alliance building, and face recognition, all played a role. I have my own views about why not just the brain, but many human attributes, such as voice, intelligence, posture and looks were selected sexually. Once hominids began lumbering across hot, open plains in a body initially designed for survival in lush forests, the esthetics of the body beautiful changed dramatically. If chimps are near kin, why does a chimp look ugly

[93] We have learned from computers that it is harder to recognize faces than play chess!

but the unrelated fawn looks beautiful? Why for women parading their beauty is it a 'catwalk' and not an 'ape-walk'? (Why are women called birds, cats, fawns, or fillies, but never chimps?) A bird can sing but a chimp cannot, so why do humans sing? Well, which body seems more suited to the crossing open plains, a bird or a chimp? If a human wants to attract mates the best way is to simply not to look or behave like a chimp. Singing, storytelling, or fluid movement would each go its own way to selecting for this end.

Sexual selection works between individuals, along behavioral and environmental adaptation. This refines adaptation in populations, but in hominid evolution populations also competed until one group emerged, so the evolution of the human brain is best understood in those terms. To see why, allow that hominid lines were competing to increase brain size for many reasons. However, when the hominid brain reached an average 1350cc further competition for brain expansion ended. Once this size was reached further evolution of all the other human characteristics also seemed to end, and there were no more migrations from Africa. Humans, with the new brain suddenly spread everywhere, displacing all the other hominid lines on Earth.

Also crucial for brain development were speech and language. To organize a tribe for hunting, combat, or general survival, speech would give any group a commanding advantage. We are not certain if *Homo sapiens* alone had speech, but more likely rival hominid groups had less advanced forms of speech. Whatever the case, unless there is evidence that speech was equally developed among the hominid groups, which is unlikely, there is almost no purpose to looking beyond speech for another trait that could provide such an advantage.

It is not to say that there is no evidence for language areas of the brain, genes for speech, PET scans of speech areas, or infants speaking instinctively early in life. Yet speech is still learned, and a human child raised by wolves could not speak, whereas a child raised by humans that could not vocalize could still emulate speech. As for brain size, it is often suggested that Neanderthals had a large brain and porpoises have a large brain. Yet brain size alone does not guarantee speech, more than the parrot's voice chords guarantee it.[94] Instead, unlike a babble of animals, any sentence, spoken, heard, or visualized, can only be actualized into meaning in the reasoning part of the brain. Language requires a degree of abstraction, and the origin of abstraction does not lie in a particular gene, but it might lie in a different effect.

Even so, I do not consider that speech alone, as mere voice organs or language modules, was decisive either. As mentioned, there are other

[94] Neanderthal man had a larger cranium than humans, but possibly a lesser packed neural density, especially for body size. It is also claimed that the Neanderthal baby was an extra three months in the womb, which would reduce the crucial window of post-natal learning.

selective advantages for a melodious voice, such as the sexual attraction of not sounding like an ape. (This can also account for why we can sing.) Nor does it seem likely that speech evolved via hard-wired reflex alone. Every new behavior in the last 400 Ma evolved in the learning cortex, so there is no precedent that this trend would suddenly reverse for speech. Especially, because speech evolved in the last 250,000 to 50,000 years,[95] the human neurology most likely to be evolving in that period was the highly homogeneous frontal cortex, the most learning-enabled, context-free part of the brain.

Early neurology evolved for reflex. Then circuits evolved for simple learning such as imprinting. Mammals evolved learning circuits that were rewired after birth, and these were able further to lower the cost of neural evolution, because one circuit design could perform many functions. Once brains could evolve quickly, across all biota the cost of evolving new novelties in any species must be measured against the cost to an ape to evolve a bigger brain. We can see this effect in flight. In any age, there is a fitness cost for a large animal adapting to flight. Until birds evolved feathers, pterosaurs could evolve wings at the least cost of change. Yet once feathers evolved, they were so efficient that long-range flight became *saturated* for other tetrapods. Mammals such as bats can adapt short-range flight, but there is one more option for flight in mammals. It is to build an airplane. Remarkably, this adaptation is fast, only five Ma of evolution from human ancestors to airplanes.

The learning neurology is "proximity mapped" to parts of the brain where it is most efficient for a function to be located. Neural circuits that require emotional input, say, will be proximity wired to the emotional centers of the brain. Even so, because most of the human brain is used for tasks that appear reflexive, it does not mean that each neural circuit of the upper brain is reflexive in a sense that each function needs a single gene or allele to express it.

I believe that out on the plains of Africa when the human frontal cortex expanded that extra few cc, something unforeseen happened in the history of life, and perhaps in the universe. Everything evolves at a cost, and the human brain has evolved not one more cc in volume or one more neuron than it needs for the survival and reproductive needs of the individual.[96] Yet statistically, brains still vary in size and neuron count, and different learning experiences can dramatically increase the number and complexity of neural connections.

For 90 billion neurons and trillions of neural connections, perhaps the human brain sits at a critical mass of neuron count to learning ratio. Below that mass, the brain is back in the plains of Africa, struggling to hunt,

---

[95]Very hard to pin an exact date, I even wonder it might be earlier.

[96] I use "plains of Africa", but it is possible that the final tribe evolved closer to the shoreline or at least along riverbanks (though nothing to do with the "aquatic ape").

survive, form alliances, and procure offspring; for typical selective reasons. Above that mass, however, the brain leaves Africa, evolution, offspring, and hunting. Like Newton, "Voyaging through strange seas of Thought alone",[97] above a critical neural mass the human brain both cuts itself off from the universe, but finally becomes a part of it.

The 90 billion neurons and the trillions of neural connections, become a critical mass of firing states of synapses that *internalize* the thinking state of the brain from the external world.[98] From internalization of the firing states comes abstraction, and higher mental activities, including speech, logic, mathematics and philosophy. Abstract thinking consumes metabolic costs, but zero genetic cost to evolve further.

This balance of the human neural mass might have a final evolutionary advantage. Humans evolved from an ape-like ancestor for a cost of 1.5% change of genes, and fusion of two chromosomes. That is dramatic efficiency. Moreover, the human species can adapt into almost any niche on Earth, or even explore outer space, for near zero change to the genes that the species migrated with from Africa.[99] That is also a dramatic efficiency. The most dramatic change though, is that the human brain can alter its state from a reflexive organ of survival to one of abstraction for zero alteration to genes. That is perhaps the maximum efficiency that the evolutionary process can achieve.

---

[97] This is quoted from Wordsworth. In earlier books, I referred to how Steven Hawking could ponder the universe from within a crippled body, but was told that this might be offensive. Still, I think that Steven's own quote from Shakespeare; "I could be bounded in a nutshell, and count myself a king of infinite space", states the principle anyway.

[98] The first cell was a form of information barrier, inside which internal order can increase against thermodynamic probability. The synaptic mass of the human brain might be a further information barrier, which allows large increases in order for little genetic change.

[99] Again, the 'genetic distance' between the original African races and that of the migrated species is very controversial and hard to measure for original populations anyway, because of subsequent mixing. Migrated species had to adapt to different, often more seasonal and more challenging environments, which produced some difference.

## 3.3 The Human Paradox

> "In the distant future I see open fields for far more important researches. Psychology will be based on a new foundation, that of the necessity of acquirement of each mental power and capacity by gradation. Light will be thrown on the origin of man and his history." **Darwin**

If life evolves in a pattern, such that complexity allows organisms to adapt for the least cost to alter genes, this process has a culmination. The end is when a species evolves that can adapt by learning, culture, and behavior, without the need to alter key sequences.

These essays were written earlier, for a slightly different topic. Even so, a challenge for a model of evolution is how it can explain evolution's most startling product.

Now, some people do not see this.

They claim that life evolved, but there is no need to explain humans. Other changes could have had a different result, but what is that point? If humans turn out to be the only intelligent species, there is no point of a theory, if this cannot explain humans. Instead, the question is that from the cultural, behavioral, or biological distinctions of humans from other animals, why from the same processes of evolution were the results so different and how different are they. To give perspective, the thesis here challenges the existing models of evolution, even how competition began in the prelife. If existing theories require revision, human evolution and behavior were an initial controversy.

In this sense, unless we can explain human evolution, there is no point, especially if we are the only species in the immediate universe, that evolved this far. It is possible, even in a fecund universe, other planets had developed highly evolved species, but these did not develop a science. For example, our species could have cut off, at ancient Rome, or at the Inca level, but not go any further. Until we have more information, we cannot say. In this, I have broken evolution into several stages. The first, second, third, and the fourth stages are for evolution, for pre-Darwinian, then early Darwinism, then the evolution of higher life, and the evolution of multi-celled organisms. The fifth stage was the evolution of thinking organisms, and this resulted in science. Until we know this, we cannot be sure.

For evolutionists, the debate illustrates three broad concerns.

The first concern is why humans evolved at all. Once evolution is accepted, it does not seem necessary to explain why snakes or leopards evolved. However, humans seem to be a unique species, "in the image of God", that exist for a purpose. Yet evolutionists themselves insist that the process is not towards a goal. I wanted to minimize quotes in this book, but in my first book I played up this dispute by juxtaposing these quotes;

> "Thus, from the war of nature, from famine and death, the most exalted object of which we are capable of conceiving, namely, the production of higher animals, directly follows." **Darwin**

> "This of course is nonsense. Evolution is something that happens to organisms. It is a directionless process that sometimes makes an animal's descendants more complicated, sometimes simpler, and sometimes changes them not at all. We are so steeped in notions of progress and self-improvement that we find it strangely hard to accept this." **Matt Ridley**

Arranged this way the quotes are unfair to both authors.[100] Darwin was trying to introduce a theory that seemed against the perceptions of that time. Matt Ridley is countering a misconception that organisms evolve to achieve greater complexity. Even so, over the history of life organisms often became more complex as a fact. Larger animals also became more intelligent and adaptable, and new species were able to evolve in shorter times for smaller changes in genomes. So, is there an evolutionary trend that leads to intelligent life?

For example, if humans discovered life on another planet, even if not intelligent, could we infer if intelligent life can evolve there? Although Earth has an intelligent species, why is there only one? Is this a rule that would apply to evolution anywhere, or is it an accident of life on Earth? Again, some evolutionists do not see a need to explain humans, but again I pose a question. If we turn out to be the only intelligent species of life in the universe (not certain yet), what is a theory that cannot explain it? Without intelligence, there is no point trying.

The next concern is the large human brain.

Even if bipedal, upright walking, tool-using hominids evolved, and many branches of them did, it still does not explain the very large brain. Hominids walked upright for almost five Ma, and lineages with a brain 40-80% the size of the human brain survived, hunted, and migrated, for long periods, without needing a bigger brain. Large brains extract high costs, especially in increased energy consumption and difficulties in the birth process. The biggest concern, however, is that the large brain seems most needed for attributes such as language, mathematics, art, or music, so it is difficult to determine why these attributes evolved. Horses have large brains for an animal of flight, so that they can flee from intelligent predators. Modern predators need large brains to catch intelligent prey. However, this still does not explain why one species of hominid needed a brain able to compose music or recite the Iliad. Of the many quotes, this one from a renowned physicist best captures the dilemma;

---

[100] Matt Ridley was not directly confuting Darwin. In my first book, the opening quotes of each chapter were arranged this way to alert readers to the issues. The quotes are fair to both authors over the issues.

> "There has been no significant biological evolution, or change of DNA, in the last ten thousand years. Thus, our intelligence, our ability to draw the correct conclusions... would have been selected for on the basis of our ability to kill certain animals for food and avoid being killed by them. It is remarkable that mental qualities that were selected for those purposes should have stood us such good stead... There is probably not much survival value in discovering a grand unified theory." **Steven Hawking**

Significantly, Darwin and Wallace, who co-founded evolution, disagreed over this issue of the unique purpose of the brain.

The final, also contentious issue, concerns the transfer of lessons from animal behavior to the study of humans, or simply, the extent to which genes determine human behavior. E O Wilson, the founder of the new study of sociobiology, posed the issue this way.

> "If our genes are inherited, and our environment is a train of physical events set in motion before we were born, how can there be a truly independent agent within the brain? ... It would appear that our freedom is only a self-delusion."

In rebutting the so-called deterministic views of E O Wilson and others, the co-founder of punctuated equilibrium theory, Steven Gould, retorted;

> "Complex organisms are not the sum of their genes, nor do genes alone build particular items of anatomy or behavior by themselves... We fall into deep error, not just harmless oversimplification, when we speak of genes "for" particular parts or behaviors."

The opposed positions that these two eminent evolutionists adopted, and why, has been discussed in many forums, including my own earlier book. Here, the only consideration should be how the lessons of such debate bear on the issue of genetic change in populations.

The lessons began with the *Origin*, and Darwin's excellent chapter on instinct. From this, it appears that most behaviors that animals enact instinctively, including courtship, aggression, cooperation, dominance, sex, or even rape, have an evolutionary origin. Moreover, many social behaviors, such as slavery in ant colonies, again first noted by Darwin, have evolutionary explanations. Darwin noted the effects by observation, but many behaviors have since been proven to lie in the genes. So, the question is that if genes can explain animal behavior, might not a similar process explain many behaviors in humans?

To be fair, in my first book I argued against this position, but since then my views have equivocated. I still disagree with claims, such as of E O Wilson, that "our freedom is only a self-delusion", and I worry how such statements play with a public not trained in scientific caution. Even if genes do play a dominant role,

For instance, suppose data showed that rape or slavery are genetic traits in humans; would this information exonerate that behavior when humans enact it? To me, this is the danger of such claims. If anything, though such scientific issues are unresolved, morally and politically the debate is in limbo. E O Wilson pleaded that "the consequences of genetic history cannot be chosen by legislatures". However, counter-arguments such as those of Steven Gould and others seemed to carry the political day. However else genes affect human behavior, they are likely to be kept separate from politics or legislation for now.

Still, all these issues of genes and human behavior have so far been framed only in terms of a standard gene-frequency model. This model applies for a population in which a gene or allele sweeps from a lower to higher distribution in one population, but the model has limitations. One is that if an allele exists at low distribution, it must have recently evolved in that species alone or the genes must at least exist in a near kin species, such as in chimpanzees. This makes typical arguments about a gene for slavery in humans, for instance, highly problematic. A gene for this trait might exist in ants, but it does not exist in chimps, any near-kin human species, or any mammals. This infers that such a gene would have to evolve just in humans, and then in a p, q *allele* form, in a way that it could sweep to fixation. As we go through the list, slavery in ants, rape in mallard ducks,[101] prostitution in humming birds, we are left wondering what all these unrelated species have to do with hominid, or even great ape behavior anyway.[102]

Of course, genes or alleles not present in chimps could still evolve uniquely in humans; but it must be in a plausible context. Only 1.5% of human genes are different from chimps, and these had to accommodate the changes that transform a great ape into a human, including the upright stance, large brain, unique human behavior, and so on. This also has to be within the time context. If behavior comes from the brain, the hominid brain only expanded in the last million years, with a final spurt in the last two hundred thousand years. Alleles can sweep in a population in twenty generations, but only if those alleles exist already, or evolve from a minor mutation. There were important changes in this short period, which also allowed for language, psychology, and a large brain. Except that once these attributes evolve, they allow other ways to explain rape, prostitution, or slavery, apart from explaining them.

---

[101] Perhaps readers are familiar, but the famous rape example was for this species of duck. How the example was noticed though, was not merely that an outside male stole in and coerced conception on the female. Rather, after this the cheated partner immediately mounted the female to inseminate her as well, which gives his sperm the best chance. It is interesting, but how applicable is this example to human behavior?

[102] To be fair, sexual behavior is always highly variable, especially in great apes. Coercion, adultery, elopement, and seeking sexual favors, are commonly observed in the wild.

For example, much has been made that some human societies, such as the Incas, had a harem-type marriage system in which the powerful alone were permitted to procure offspring. This is offered as proof of a genetic impulse because many species, such as seals, also use a harem system. Another suggestion is that not only is the harem system natural for humans, but that monogamous pair-bonding is an unnatural system, imposed by authorities such as the Christian Church to win popularity with lower males. Such theories are not only historically doubtful, but they do not explain the genetic mechanism.[103] For instance, if a gene for harems is present only in populations using that practice, the gene must have evolved suddenly, because harem-based human societies have not lasted long historically. By contrast, if the harem gene is for the entire human race, it must revert to a bottleneck on the plains of Africa. Except that this is a problem, for all species-wide genes for attributes in post-primitive society, such as harems, social classes, or slavery. These social practices did not exist when the human population was concentrated in a way that a single allele could easily sweep through it.

This brings us back to the problem of the fitness conditions that will enable any gene to fix. E O Wilson argued that; "societies that decline because of a genetic propensity of its members to generate competitively weaker cultures will be replaced by those more appropriately endowed". Well, no one could agree more, except that if it is true it confutes almost all the other arguments. There is no rule on this, but as a trend societies that are egalitarian, with say, a marriage system based on free choice in partners, have proven stronger.[104] Especially so when these societies collide with highly structured ones, based on a harem marriage system, or a slave-based society versus a free citizenry, at least in modern times. Yet the puzzling aspect is the presumed fitness model of how genes for harems, slavery, or feudal-type social structures are supposed to spread. It seems to repeat the error of Lamarck; that just because an influence arises, the biology tracks the volition. Why in Darwinism should a gene for slavery spread in a society that imposed slavery as a structure from the top, or for any comparable example?

Perhaps the point is labored, especially as the issues have now been debated many times by eminent evolutionists who weighed in on both sides. Even so, issues raised in a standard context have not been solved in that way. People will plead how convinced they are that explanations of human behavior must be Darwinian, but often explanations veer towards Lamarckian. Richard Dawkins, after everything that he proposed about genes being selfish, suddenly turned face for human behavior.

---

[103] I am not aware that Roman Law supported a harem system, when Christianity began to spread, or that early or isolated human tribes particularly supported harem systems.

[104] Males will work and fight hardest for their own children; not someone else's.

> "As an enthusiastic Darwinian, I have been dissatisfied with explanations which my fellow enthusiasts have offered for human behavior. They have tried to look for 'biological advantages' in various attributes of human civilization… for an understanding of the evolution of modern man, we must begin by throwing out the gene as the sole basis of our ideas on evolution."

R Dawkins then proposed that non-biological *memes* were the units of selection in human behavior, which allowed him to use natural selection at all. This is the dilemma. While there is much competition in human society, it is hardly for genetic fitness, such as passing on one's DNA. Even sexual conquest is hardly for the end of passing on DNA, as many people try to avoid offspring, or use conquest for non-reproductive forms of sex. There are many such conundrums.

Ironically, the answer might be one more quote by E O Wilson, in which he claimed that his theories could only be confuted if there were discovered "a new and as yet unimagined form" of genetic change in populations, which I think there is. While the standard gene-frequency method is powerful, its fitness model is restricted to contexts that are difficult to generalize, for instance, about human motives. However, if genes compete not just to copy in numbers, but also to conserve sequence as they spread, it brings a fresh perspective.

One perspective the new model might bring is mitigating some of the controversies surrounding evolution theory. In this, evolution is often criticized due to difficulties of the model, in the areas that this book has discussed, such as the evolution of sex, or the large transitions of life. A challenge is then issued by critics of evolution to explain love, religious belief, or self-sacrifice in humans, via a model that is already incomplete in other areas. However, the standard gene-frequency model was only to explain how favored alleles distribute in single populations. This was a major problem for evolutionary theory when it was first raised. That the standard model solved such problems shows the robustness of the approach. Nonetheless, to try to explain every nuance of human behavior (such as why people believe in God) via this model, can often result in explanations that seem desperate.

By contrast, the model here teaches that genes are competing not just to gain copies, but are competing to conserve sequence as they distribute. This expands the scope of any explanation. For instance, an advantage might arise from why a gene spreads, but it can also arise from how a gene retains its sequence. In this sense, the advantage to learning, culture, or psychology-based behavior, is that humans can use these to adapt for many situations, without core gene needing to alter. It is why, for human adaptation, we should look for a balance, between a direct gain for a favored allele to spread, and an indirect gain for genes to avoid altering sequence to adapt.

Another challenge is to identify genes that allow large changes in the organism (*phenotype*) for small changes in genes (*genotype*), and the next diagram (Fig 3.1.1) gives an example of how this could apply for human brain size. Note that while adult human and chimp skulls look radically different, infant skull shapes are similar. The human skull looks more advanced because it stays closer to the infant skull shape, an effect called *neoteny*. The real difference though, is not neoteny, but that such changes would be simple genetically, because they do not require new genes. Instead, the changes in brain shape are geometrical.[105] Moreover, genes controlling how brains grow, such as the BF-1/2, existed for millions of years, so brain size was another means for these ancient and highly conserved genes to spread. By contrast, to explain why the human brain is large based only on how modern alleles spread (a "peacocks' tail" or similar story) can be convoluted. However, if a large brain could evolve from small changes in the genes from a chimp brain, such convoluted explanations are no longer needed.

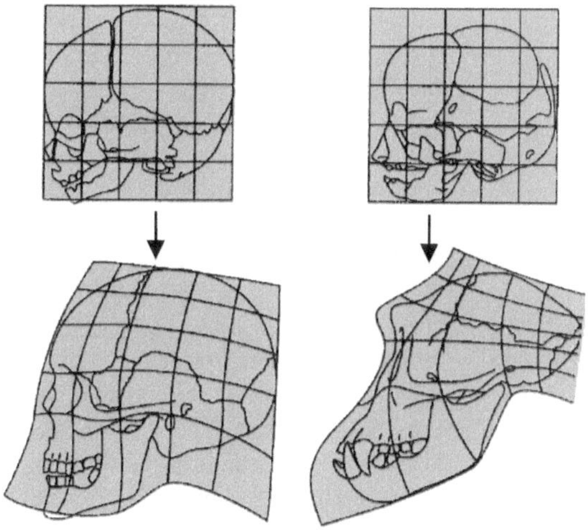

Fig 3.3.1 Developmental pathways of human and chimp skulls, from infant (top) to adult (below). Such radical changes might be possible for very small changes in genetic distance in the genes controlling change.

Essentially, humans evolved in ways that maximized the options of adaptation, for little change to core sequences, and only a small genetic change from chimp-like ancestors. This way, humans can adapt to every

[105] This is adapted from John Allman's book, *Evolving Brains*.

continent on Earth, and express a huge range of behaviors, with very few core genes needing to alter sequence. Humans are sexually reproducing, so alleles are reshuffled each generation, which changes genome DNA. Humans also are modern eukarya organisms, so they also contain junk and highly variable DNA. For all their behavioral and racial diversity, though, humans have a more homogeneous genome than chimps, despite that humans are more widely dispersed.[106]

Another illustration occurs when many features have to alter in conjunction, such as in the evolution of mammals, especially, the higher metabolic rate than dinosaurs. Moreover, their rate of evolution was not linear in time, but related to the density of mutation-selection events. Dinosaurs and mammals began an evolutionary journey from similar points. Yet over the same time, the more thermodynamically probable dinosaur could afford to radiate widely. The mammal though, could never have evolved so far in the same time in a distended way. It took several orders more of mutation-selection events for the mammal to evolve, and this was achieved not by radiation, but by a tight, focused, form of evolution in a narrow line.

Of course, species do not do a calculation of the thermodynamic cost to evolve each attribute; they compete blindly for food and reproduction. Yet the conditions of life, forces each one to compete under different conditions. High metabolic rate organisms have to compete for more food, but often for only reduced clutch or litter sizes, so the attributes of high metabolism need an impetus to evolve. To stay active during the cold mammals did not only evolve hotter blood, but endothermic regulation, body fur, and a gullet for breathing while swallowing. Unlike for a longer leg or sharper claw, such attributes take time to evolve and to work together.

In real life too, organisms do not foresee a total cost of change; each organism merely struggles to survive and reproduce as best it can. As Richard Dawkins has said, the process is blind. By contrast, engineers in large-scale technology often try to forecast the cost of change, such as when facing the dilemma of whether to continue upgrading an outdated product, or to replace it with a radically new design.[107] Natural selection however, working blind, seems able to forecast with a similar efficiency. During the Mesozoic say, natural selection kept the pterosaurs flying. Yet hidden away it was able to evolve a radical new flying animal, the bird.

---

[106] This is very contentious, and might not be true. Humans apparently vary widely in non-coding DNA, which plays a more important role than first thought. There are also different levels of genetic diversity between original African populations, and groups that migrated from Africa, although in modern times this has become intermixed again.

[107] The classic example was during WWII, with the issue of whether to keep upgrading prop planes, or to build a jet. The solution was to keep upgrading the prop planes, but set one team aside to work on the jet.

Evolutionists think at how the need of the gene to spread its sequence resulted in the beautiful productions of nature.

Next, consider the problem of identifying the fitness conditions under which the gene is supposed to spread.

Remember that in standard theory the gene benefits only itself. Two alleles, "$x_1$", "$x_2$" will be competing with a conferred fitness $w_1$, $w_2$ about a locus *X*, and all other genes and loci are ignored. (Roughly, "$x_2$" will spread more than "$x_1$" in proportion $w_1 < w_2$.) For humans, certain alleles such as for sickle cell disorder are known to confer fitness and distribute this way.[108] Yet how does one quantify the fitness value, for example, for rape? The human female's ovulation is concealed, so the $w_i$ for rape is must be about 0.2 for a random conception. There is also a chance that the rapist could be killed by rival males, or that a lone or rejected female could die in childbirth. By contrast, a man who lives in a family unit where both partners are fertile is almost guaranteed reproduction. It does not mean that alleles for rape, or many violent or coercive behaviors will not exist, distributed over the human genome. They exists, yet to claim that rape exists in human society solely because about a locus *X*, an allele for rape has a $w_i$ higher than an allele for non-rape is not computable.[109] The fitness to rape might be computable, say, at a locus for mallard ducks, but not in a different population.

Another concern is the concept of the single population. In this case, it means a collection of individuals in which a gene could obtain 100% distribution. If a species is a thousand or less individuals in a connected area, a favorable allele has a good chance to distribute across it. On the other hand, if the population is large, millions of individuals that may be scattered over the planet, the groups are like its own population.

This also happens a lot for humans, where genes and alleles for race will often dominate just one group.[110] Scattered populations like this also modify statistically, where an improved mutation might occur randomly across the different groupings, but slowly fix. Large populations can be devastated by fatal disease, and new disease resistant individuals replace them. The issue instead, is when a researcher generalizes that a human gene evolved "for" a particular behavior. Often, there is little specificity, about which population mechanism is involved.

In summary, a model of gene distribution in a single population becomes overly challenged if trying to explain the complexities of human evolution and behavior.

---

[108]In a famous example, sickle cell disorder offers certain immunity against malaria.

[109] Whether a rapist will be killed is almost impossible to quantify. Unlike for ducks or even chimps, the human male can kill rivals. Rape is often in gangs, except that then one needs another gene to explain the gangs, and who gets first insemination, and so on.

[110] Say, populations adapted to live in high mountains, such as the Sherpa, have undergone many modifications of the breathing and circulatory systems to live in high, cold places.

Instead, to approach this from a standpoint of genetic evolution, one should consider the distribution of genes, including how the successful organisms will adapt for the least cost of change in the conditions of struggle. Humans evolved rapidly, and changed greatly from chimp-like ancestors, for just a 1-2% change in their genes. This small genetic cost of change reflects why humans have attributes such as culture, language, or moral inhibition. Attributes like these depend on cooperation to work, which is puzzling why genes facilitated these to evolve. However, such attributes make organisms adaptable for little change to core genes. It allowed humans to multiply in huge numbers for slight changes in gene sequences. The evolution of any human attribute must also be seen in a context of the general saturation of life's large animal possibilities, where further adaptation had to be radical in its results, for a minimal genetic cost to alter key genes.[111]

Next, consider in terms of genes struggling to conserve sequence as they distribute, why the human species is moral.

[111] There were no radical changes of genes, but again, human DNA did vary greatly and more rapidly in human evolution than had been estimated when this was first written.

## 3.4 Evolving Moral Constraint

Why do humans exhibit moral behavior?

As mentioned, apart from why humans evolved, or the large human brain, the next contentious issue is the extent to which genes determine human behavior. This is a huge topic, but it can be crystallized from the issue of why humans have morals.

Briefly, no idea is simpler than genes coding for characteristics that enable the genes to spread. Ideals associated with morality; selflessness, cooperation, self-sacrifice, or sexual restraint, however, run counter to how genes maximize distribution. So, why would genes for inhibition against greater distribution evolve in the first place?

In this, evolution of morality is a crisis of the standard model. Other limitations to explain the chromosome, sex, the fitness landscape, or rogue DNA, start from first life. However, there is always a hope that a gene will be found that sweeps a new path, that can explain everything. For morality, though, this hope has nowhere left. There is no reason why a gene would evolve for its own restraint.

One answer is that there was never one gene or allele, responsible for why humans have other morally enabling attributes. Rather, humans evolved a range of attributes for many reasons. In any social and cultural context, the attributes pre-adapt a moral-enabling behavior. However, why in one social context a group of individuals would hold moral or ethical values that might be different from another group is a huge topic. Debate on this intensified because of books such as *The Selfish Gene*, and the issues go on forever.

We can also ask if the model of how genes distribute is complete.

Strangely, it was pondering this that led me to the model in this book. The conundrum is the apparent suicide of a gene. The question of why a soldier would sacrifice himself in war, for instance, can be likened to why a gene would commit suicide. It is not difficult from this to wonder if the gene might not be dying, but might propagate via a *radiating* distribution, if we knew the mechanism. The earlier chapters explained how genes could distribute via a radiating (or 'imaginary') component, if the gene's sequence distributed across many populations, but the model was for a single population.

However, even this model cannot solve all the puzzles of morality. The reason is that any *radiating* component of gene distribution would be strongest for ancient genes that evolved long before humans. As a result, genes and alleles in humans that expressed morality would still seem to evolve for their own restraint, but a restraint to conserve ancient genes unaltered in sequence. So, why would modern alleles restrict distribution to conserve the sequence of ancient genes?

Let me answer this genetic conundrum first.

The crucial point, is that long before humans evolved, throughout the history of life, each time that a new level of complexity evolved, newer genes came into existence. This was in ways that would allow ancient genes to spread into new niches, but without the ancient sequences being forced to alter to adapt. This is why the first chromosome formed, it is why eukarya evolved, it is why sex evolved, and it is why higher behavior and leaning evolved. After four billion years of evolution, this too is why morality evolved.

The impetus is that every gene that exists tries to copy its sequence into every niche of life,[112] but not every gene can achieve this exemplar. Even highly conserved genes cannot spread across every niche of life unaided, so at each phase of evolution selection arrives at a balance. Ancient, highly conserved genes lay down the pathways along which newer forms of life adapt. Complementing this, newly evolved, variable genes come into existence to expand basic designs into a multiplicity of newer forms. Every transition of life is an extension of this pattern; established attributes evolved in the past are conserved, while new variations are added.[113]

Morally enabled behavior evolves from this balance, but a balance that follows billions of years of prior evolution. By the time that humans evolved, margins of how further this balance could be pushed already were stretched. To adapt to the plains of Africa, thousands of animal species had achieved this long before the first hominids attempted it. Even for hominids, there was more than one lineage competing to adapt bipedal, primate life to the plains, and of all the lineages competing, only the one leading to humans survived. Conserved genes common in all hominids would survive no matter which lines went extinct, but for those genes expressing subtle shades of adaptation, such as moral behavior, the stakes were high. In a rapidly altering environment with diminishing resources, he who adapts first adapts best, or adapts only. Fast adaptation in turn requires small changes in genes for large changes in behavior, and this favors morally adapted behavior.

For instance, one advantage of morality is that it is mostly learned, and learned behavior requires smaller changes to genes to evolve. About 80% of the human brain is for post-natal learning, and this allows a much larger brain for a similar 20,000 genes that express a chimp brain. This way, learning allows higher general intelligence for a similar density of genetic instructions, by allowing rapid expansion of brain size by filling brain

[112] This is only a way to model a complex set of interactions, to give a result that emulates the molecular distributions that we observe.

[113] With sex for example, founding genes of eukarya are 100% passed on each reproduction into every form of life that reproduces via sex. Yet via sex too, many variations are passed on only 50% from each parent, to provide endless permutations on the variety to adapt.

volumes with repeats of similar circuit design. In addition, learning allows flexible behavior, if conditions change.

To give an example, consider the moralistic dilemma of caring for the old, sick, or wounded. When times are good this can be a moral act with long-term benefits.[114] Yet when times became difficult, if the old and sick became a burden, they can be abandoned. Similarly, the moral reasoning that teaches that we should care for the old and sick can also be abandoned and the neural space can be turned to other tasks of survival. Now, in a struggle to adapt there would be huge advantages to a species that could be moralistic in good times, but brutal in difficult times. A species like this would be more adaptable than a species forced by neural inflexibility into fixed responses (but see footnote).[115]

Notice though, that to evolve the genes to program this complicated response to a varying situation via reflex would be impossible, especially under the intense competition and rapid evolution of competing hominid species. Rather than evolve the reflex, it is simpler to enable moralistic-type decision making, for a range of choices. Yet notice also, that once behavior becomes morally enabled, behaviors can alter to circumstances without the need for genes to alter sequence. In this model, not needing to alter sequence is an advantage to all genes. A cost will be paid to adapt into morally enabled behavior, which might be reduced numbers of total offspring for each human individual (for example), but this is part of the total balance of how all of life evolves.

The more debated issue of morality though, is not the model of how genes distribute, but that humans still experience deep passions, and once civilization evolves, philosophers label these passions "moral feelings". These feelings are of love, shame, or remorse. To the extent that any feelings affect humans physically, they must have arisen by evolution, regardless of which moral needs such passions serve in civilization. The dilemma is to explain why these feelings are fit.

My view is that *feelings* are just that; they are deep physiological motives that all higher animals experience, and which evolved for many reasons. For example, primitive human behavior requires the feeling of "inhibition" for many things, such as sharing food, cooperating on a hunt, raising orphaned children, caring for the sick, or trusting women at home to stay faithful. The evolutionary dilemma is to discover why this would be fit in an individual. Well, the emotion "inhibition" (do not get yourself killed) has a simple explanation of why it is fit. The dilemma instead is that complex human behaviors process *transactions*, which involve a range of thoughts and responses that can include "inhibition". Returning to the model, some evolutionists feel compelled to explain any inhibition,

---

[114] One theory is that old people recall how the tribe survived different perils in the past.

[115] Correctly, individuals were fit if they had large, flexible brains. Groups of fit individuals with such brains survived when groups of smaller-brained individuals perished.

such as against adultery, by explaining a gene for the full transaction that inhibits against adultery, but this is nonsense. The feeling "inhibition" (do not be killed) would have evolved hundreds of millions of years ago, in a different part of the brain to the part that warns against committing adultery. Despite this, another part of the brain can reason that adultery might be safe in another town or country.

This distinction between *transactions* in the reasoning part of the brain in humans, and deep feelings of anger, desire, or fear, that evolved in different parts of the brain for different reasons, vastly complicates the approach to evolutionary explanations. Higher animals need a repertoire of moods and emotions as motivation, but an efficient design will allow these motivator circuits to be used in many responses. For example, the 'cold' sensation when people are emotionally isolated is the same 'cold' response circuit for when the temperature drops. It is easy to explain why that primal response to 'cold' was fit. Once this 'cold' circuit had evolved, it would also be fit to utilize it for the 'lonely' feeling, rather than evolve a new sensation for that feeling. Each step in the evolution of each feeling was fit, but not just as a single transaction.

Consider the notorious puzzle of so-called sexual inhibition among humans. Hominids reproduce by sex and judging by the high rates at which humans repeatedly overpopulate, they seem to be good at it. Yet, because of certain social customs, such as inhibitions about nakedness or promiscuity, humans have been stereotyped as naturally inhibited about sex. As if to prove it, many people do feel genuine 'guilt' when they violate sexual codes that society dictates. This is the puzzle. It is easy to claim that society or the church makes the rules, but deep feelings that affect the physiology such as embarrassment, flushing, or racing heart must have resulted from evolution. The question comes back; if humans reproduce and alleles spread via engaging in sex, why would an allele spread that inhibited against sex.

When trying to answer it, appreciate that this question is convoluted from the start.

Again, there is no real evidence that humans are naturally inhibited about sex, or that inhibitions against incest, or mating with an unreliable partner, do not have other explanations. All animals have inhibitions regarding sex; to avoid a diseased partner, against sex with other species, or with a relative, and so on. This is easy to explain. The conundrum with humans is that many inhibitions are psychological, so that not all people will feel bound by them. Suppose that a male had no inhibitions about cheating. Would he not have more illicit affairs, and hence more offspring, so his genes would spread? This is the old quandary about sex, back in a convoluted form.

This is two questions. One is; what are the advantages in technology of switching from hardwiring to soft wiring? The second question is; will not soft-wiring make it easier to "cheat"?

The answer is that soft wiring is easy to reprogram and is easier to adapt than reflex. In regards to cheating,[116] one solution is to leave safety circuits hardwired, but to soft-wire responses that need reprogramming. For instance, cheating will involve not one, but several transactions over many parts of the brain (Fig 3.4.1). Some responses will be in the lower, primal, hardwired part of the brain. Some will be in the emotional seats of the middle brain. Whereas planning a tryst with a partner will be in the autonomous, pre-frontal cortex.

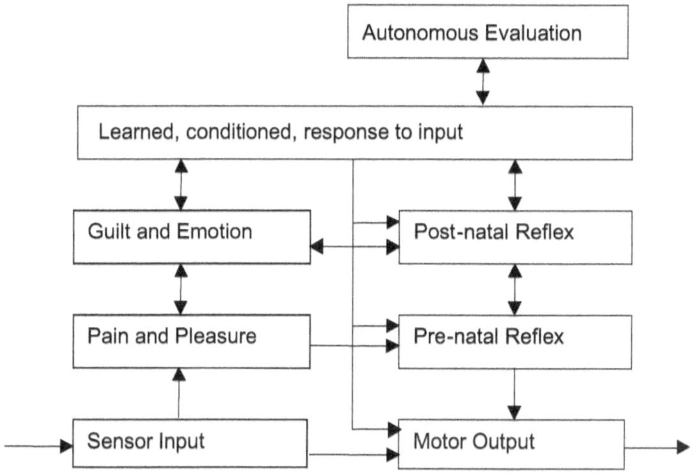


Fig 3.4.1 Whether for hunting or evaluating morals, the human brain is complex. We need to establish why it was fit to evolve such a complex brain, before asking why feelings of moral inhibition might arise in such a brain. Complex brains will feel "mood and emotion" but from a mix of inputs, from many parts of the brain.

Whatever the transaction though, if it translates into more offspring, genes for that behavior will increase frequency. Generally, an individual with all the facilities of his or her brain fully developed will be fitter than an individual with crippled emotional responses, or an individual unable to think situations through logically.

Research supports this contention. An individual can have damage to the emotional, moral-enabling parts of his or her brain, so we wonder if such people might not be held back by emotion or inhibition. However, loss of inhibition or emotion does not make anyone more logical. On the

[116] I discuss cheating and how moral or ethical precepts arise more in my first book.

contrary, such individuals can become either paralyzed with indecision, or erratic and impetuous, neither of which is fit. Instead, to be capable a brain must be balanced over all its facilities. We might suppose that morality is a contrast between cheating and honesty, in which cheats will prosper. However, in a computational sense, morality can be likened to an ability to resolve choices by judgment or principle when facing the unexpected, which is fit behavior.[117]

Neurology also works on a basis of proximity.[118] Reasons for an inhibition need not be directly hard-wired by reflex, but the learning neurology for that function in the upper brain can be routed close to the inhibitory responses in the middle brain. This enables regulating moods to condition learning or inhibit behavior. Such inhibition would be fit for the purpose for which it was originally adapted, although it might appear as a moral inhibition when we try to explain it.

The human brain also contains a ratio of 75 billion "learning" to 15 billion "fixed" neural circuits. Unless there are other controls for such massively flexible neurology, the brain will not process in an optimum way. This is why evolutionary explanations of morality should show first why a brain with a high ratio of learning to reflex is fit. Only then should the focus shift to why attributes such as feelings contribute to fitness, although these might have been pre-adapted for other reasons. For humans too, the extended birth-growth cycle locks the species into long periods of nurturing the young, which further requires circuits to ensure the power of learning over reflex. With the first 15-20 years in a lifetime used in learning, humans only obtain evolutionary payback for prolonged learning if it is passed on. If parental instructions are passed on in genes, retaining them is automatic. Within a human eight-to-one learning ratio, only 15% of instructions will be 'unbreakable' genetic code and the rest have to be learned. That is why if learning is to be effective parental instructions have to be accepted.

Moreover, while we might suppose that morals, religious feelings or inhibition derive from learning or culture, it is not learning or culture that make these feelings so powerful. We can think any thoughts that we wish in the rational cortex, but like a computer, the highest part of the brain is without feeling. Feelings instead originate deep within the lower brain. The sentience that motivates lower animals to escape pain or seek mates exists in humans, but it is also wired into the rational thought processes of the upper brain. These feelings are powerful. Ideas can hurt. Memories can sear like heat. Associations can bring ecstasy.

---

[117] A robot probe to another planet would be more valuable if it did not just respond in a fixed way, but could reassess mission goals if facing the unexpected, based on principle.

[118] This is 'proximity mapping'. It means that the decision-making is kept in proximity to emotional response, and is reinforced by it. It allows that the brain be highly structured, but leaves a flexible circuit design, that can be modified as post-natal experience.

"For love *is* strong as death. Jealousy *is* cruel as the grave: the coals thereof *are* coals of fire, *which hath* a most vehement flame." (Song of Songs)

If humans feel shame, love, or grief, the sentience is a product of reflex, and hence evolution. Yet why humans feel any particular effect arises from perceptions in the higher cortex. The brain has optimized the moral response function until it becomes the most powerful reflex, overcoming fear of death. Yet these powerful feelings can also be tapped by the insentient thought processes of the upper brain.

Even if it is possible to explain anger or desire this way, though, what is the evolutionary source of the feeling 'guilt'?

I suggest that the 'guilt feeling' can assist in rapid brain expansion. Learning neurology allows the brain to expand rapidly, but this will only work if learning can sublimate the power of reflex. Instead, if it is fit to have a large brain, it will be fit to transfer responses away from reflex to the more flexible learning circuits. The downside is that learning might not be reliable under stress, and we see this all the time. With food, reflex is to consume it immediately. The learned response is to store or share it, which offers long-term options, but needs often conflict. In animals (like squirrels storing nuts) needs are embedded in reflex, but this sacrifices behavioral flexibility. The other way to enhance learning is by evolving a special reflex; that we can call a "reinforcement circuit". Such circuits would ensure that a transfer of reflex into leaning rewards the individual with 'good feelings' (positive reinforcement) when learned responses are retained, but these will produce 'bad feelings' (negative reinforcement) for failures to abide by learning.

For example, there might be two motivations: one to eat food, one to preserve it. The first instinct will be genetic, but if the need to preserve food can be converted into a learned response, its length of delay can be varied based on needs, which would increase the options. Hunger, which is reflex, is always there, so another feeling must counterbalance it. If humans can override reflexive drives by learned ones, and the constraint holds, we feel good inside. If we attempt to countenance reflex and it fails, in that the learned behavior does not prevail, we feel bad inside. If it were fit to convert constraints to learned ones (for a large brain), it would be fit to evolve a reinforcement mechanism so that learned constraints will "hold" as effectively as reflexive ones.

Not just feelings, but also many human attributes that enable moral-appearing behavior evolved for many reasons. For instance, a large brain allows moral reasoning, but it also assists hunting and foraging, which is directly fit. Similarly, the human face has a versatility to express shades of judgment, such as the silent rebuke that 'this disgusts me'. Yet, while human facial language can be used for enforcing group values, the value can be about many things; sex in one instance, but courage in battle for another. So again, the fitness advantage to facial expressions need not be explained solely as the moral values that facial contortions express. While facial expressions are an attribute of human behavior, chimps evolved complex facial signaling too. Whatever the fitness advantage it existed for at least five Ma, long before moral concepts in an ethical sense first arose.

If anything, all higher mammals that have social behaviors such as group loyalty, cooperation, and conformity in the herd or pack, also have morally enabling attributes. That is why it is possible to train a dog to exhibit behaviors, which are moral in human terms, (such as sacrificing the dog's life to save its master). Dogs can be taught such behaviors while a cat might not, because of how dogs once survived in the wolf pack. Even then, one need only explain the behaviors of dogs in natural terms, not moral ones. In this way, moral behaviors occur for fitness reasons, but not direct ones. Humans have a large, learning capable brain, which easily connotes moral ideas. As well, humans evolved communications that powerfully influence group behavior. These include voice inflection, versatile facial expressions, plus head, body, hand, and arm gesturing. Our species also has a group-oriented biology in that humans hunt and live cooperatively, provide the female assistance with birth, and jointly nurture the young for long periods.

In an ethical sense, morality is a social code that civilized humans generate, but the basis for this code is moral "feelings" that all humans experience. The source of these feelings has many explanations, but the primary one is as a means to maximize behavioral flexibility and rapid brain expansion, for a minimal cost to alter and modify core genes. Other organisms remain trapped into the behaviors that they inherit.

Morality is the crucial attribute that frees humans from this limitation, by selecting them to control behavior outside of impulse.

In summary, debate over the evolutionary sources of morality began from a claim that human morality was only an illusion of the need of the gene to maximize spread. Apart from other limitations, the issue was that the standard model for gene distribution was beset by paradox, including to explain how the model evolved. This was the conundrum. A model that could allegedly explain morality after billions of years, could not have worked anyway, before other attributes of life needed for its modern expression evolved. Instead, if we reconsider how molecular competition likely arose, then continued down the ages across the transitions of life, other explanations of morality become possible.

If anything, it is not until we realize that every gene is also trying to conserve its sequence as it distributes, that morality starts to make sense. The need for moral-type decision making is encoded in genes, but this encoding evolves because when the only constraints left on behavior are moral ones, no genes need to alter their sequence further for that lineage to adapt.

This digression into the causes of morality ends this evolutionary journey. It began *pre*-biotic via a competition not for an explicit biotic end, but for early molecules to self-replicate with greater copy fidelity than rival molecules. The journey ends four billion years later, slightly *post*-biotic, when a species evolves that can adapt at maximum rate via learning and culture, outside of a biotic need of core genes to alter to adapt. The journey also ends when a brain evolves to a critical neural mass sufficient to isolate thinking from reflex. Isolation of thinking from reflex maximizes options of behavior, once sufficient moral regulators of behavior evolve.

After a billion years of reflex, the journey of higher behavior ends when a species can maximize all its options of adaptation, and the only constraints left on behavior are moral ones.[119]

[119] Again, there are many conundrums of ethics, religion, culture and moral philosophy concerned with this topic that I am certainly not trying to avoid. They are discussed in my first book, but it would be inappropriate to venture too far into those topics here.

# 4.0 Summary of the Main Points

## 4.1 The Five Stages of Life

This book, having explained how life evolved on Earth, now will recast this to how life might have evolved elsewhere. However, the emphasis is still on planet Earth.

Life evolving in general this way is a new theory. Now, if there is no mathematics, anyone can have a theory, based on facts. Here, my five stages of evolution reflect how life evolved on Earth. I kept this to five stages, because there were key changes, on Earth, we must mention. I could extend this to six stages, or many stages, but five is a number, with no point to go too far.

Oddly, the key stages are Stage 3, Stage 4, and Stage 5, because there are problems here. Over the universe, Stage 1 will have occurred, wherever there was water and the right conditions. It might even occur in this Solar System, on early Mars, Europa, or other areas. Stage 1 is from natural selection, but without life as Darwinian evolution. Stage 2 follows as Darwinian or cellular selection. On Earth, Stage 2 involved two types of cell wall surroundings, for ether linkages as pre-archaea, or ester linkages as pre-bacteria. We cannot know that one, both, or extra linkages were used elsewhere.

Beyond Stage 2, though, was Stage 3, but this is where the problems start. The issue is that we do not fully know why life evolves. Up until now, it was easy to say it was by natural selection, but my theory shows that there were two types of selection. Now, molecular selection starts primitive molecules replicating, and we can be certain why it occurred. Stage 2 was the start of cellular, Darwinian evolution. Without knowing much else, we assume that this was simple. If ether linkages are the most likely, then at least archaea cells will emerge. If there are ester linkages, then bacteria could emerge. There might be other types of pre-cells, but experts would need to comment.

To come to Stage 3 though, is more problematic.

On Earth, repetition of Stage 2 was not enough. The replication where genes could clone, almost endlessly, had to be overtaken by a type, where repetition was substituted at a new level. Here, organisms increased about 1000 times, circular chromosomes were overtaken by horizontal ones, and new organisms were born, lived, and died. This was before we considered sex. Some extra factor required this new organization of higher life, which must be explained. On Earth, archaea and bacteria produced lower life, and higher life came as eukarya. We can be certain that Stage 1 would lead to Stage 2, but we cannot be certain if Stage 2 would include archaea and bacteria, or a third type, or more. We also cannot be sure that eukarya would evolve in other places, or that archaea and bacteria alone would lead to it. For this effect as well, we simply must know more than we now have about early life.

Once we reach Stage 3 of evolution, if it happened, we must decide on the evolution of sex. In life on Earth, sex was very important. From the evolution of eukarya, life could have continued for a billion years at least without adding sex. On Earth, I am personally happy that we have sex (though not everyone sees it that way). Sex allowed new forms, about six major and three minor clades to life, which allowed diversity. Sex has caused the chromosome to double again. Otherwise, it does not make sense why this occurred, within overall evolution. Sex was important, but we still cannot be sure if other planets evolved sex, or if on other planets eukarya evolved from archaea and bacteria. Ultimately, in my opinion, I do not include the evolution of sex as central to Stage 3, but instead I have this completed, with animals, in Stage 4.

To me, Stage 4 of life was the evolution of sex in animals, and on Earth, this was at 620 million years ago. Now, in the history of life, we must allow that life spread across the planet. This could have been from the evolution of multi-celled organisms, from about a billion years ago. This might be a good point to change to Stage 4, but again, my choice, is to leave it to sex in animals. With no other precedent, we want life on other planets to resemble Earth, for all that we can tell. Sex in animals will introduce a special mobility to life, not there before. We do not know about other places, but here on Earth, animals have 100–150 cell types, against the 10–20 in plants, fungi, and protoctists. Without knowing, a widespread variation of everything apart from animals is interesting, but only animals can provide for intelligent life.

If you take sex in animals to Stage 4, apart from intelligence, all the forms of evolution that most interest us are in Stage 4. Without Stage 4, we will not see evolution in action from far away, such as the spreading of life. All the major changes on Earth, such as the rise and extinction of the dinosaurs would be in Stage 4. We could suppose, on another planet that Stage 3 forms could spread across the planet. We could also assume that a planet evolved without sex, but animals, or even intelligent beings, could not evolve. In this case, we would have to modify the scheme here. Before we discover other planets to that degree, was some efficiency in evolution the cause of sex on Earth? Without knowing the other causes, it is hard to imagine a species that could reproduce, without sex, or some other restraint that limited the total output.

Instead, let us examine the stages of life, which led to Stage 5, which is the evolution of intelligence. We must know for sure, the chances of this evolving elsewhere in the universe.

First, let me repeat the main results.

Without knowing anything else, from the theory just outlined, in the universe, Stage 1 or life might be common. It could be, or could have occurred, elsewhere in our Solar System. You need water, and the right conditions. If amino acids or other molecules of life are common, I do not see why Stage 1 of life would not occur, widespread in the universe. The next, Stage 2 of life, could also be common, and if not elsewhere in this Solar System, at least on other planets.

If all that life requires to start is natural selection and water, Stage 1 could lead to Stage 2. If ether linkages are the most likely, again, archaea cells will emerge. If there were other linkages, then these would emerge. If natural selection gave rise to molecular selection for Stage 1, it would not require much to come to Stage 2. Unlike for Stage 1, which likely absorbed heat, Stage 2 would likely radiate heat. This would be the shift to life, if the conditions were right.

Again, Stage 3 would follow Stage 2, but now we must know things, and cannot just work from the math. It seems logical if Stage 2 was there, a Stage 3 would develop, but again, we must have lower forms combine, and organisms that now are born, live, and die. Natural selection will take life to Stage 2, but why does life develop further? There must be a combination, which is from a gain of energy. After Stage 3, we progress to Stage 4. Again, I include the evolution of sex in animals. If you check Chapter 2.7, there was a genetic change to allow sex in animals. Only a modification to the GK-PID genes caused sex in animals, as Metazoa, after the next lowest, Choanoflagellates. We can imagine, in some world, where lower types, say to fungi, diversified and covered the world. Yet although that is interesting, it does not lead to animals, with increased cell types and mobility, how we imagine. Again, we are stuck, and no amount of math alone can show us how it evolved.

Now, consider Stage 5.

If everything else on another planet was very close to Earth, so that somewhere it had gone through all the Stages, 1, 2, 3 and 4, what are the chances of life evolving intelligence, such that we would notice, say, by radios beacons or another effect. If we figure out that the chances are good, then where are the others? This is a very loaded question. Proof of another civilization could be present tomorrow, or as I write. Except, assume for now there is no proof of another planet, then where are all of these? This is a question, which must have an answer.

My view is that although 4.6 billion years for life seems long, in a universe that is 13.8 billion years old, it still might be a short time over the universe. Life on Earth could last another billion years. The scope for life evolving overall could be a trillion years. In this case, Earth is young, against a universe that could be much older. If nothing shows up, in say, another 50 years, we are just too early.

The other problem is that it might be difficult for life to evolve onto advanced intelligence. I think that Stage 1 of life is very common, and that a follow-on Stage 2 is common. However, it seems harder to evolve to Stage 3, and then onto Stage 4. Again, there must be a common motive for life to evolve. Unless someone can state eukarya will evolve, and then sex evolves, we cannot know that it will. Again, we can assume that Stage 4 exists, but from Stage 4, spread across another planet, what are the changes of Stage 5 evolving?

If we look across the universe, so far, we have seen no evidence of advanced life. Again, Stage 1 is common, and Stage 2 is common. Stage 3 depends on the evolution of eukarya, and Stage 4 on sex to spread in animals. I cannot say how common is Stage 3 and 4, but in the universe, it might arise. Yet, we still have no Stage 5 that we can recognize. Either Stage 5 civilizations exist, but they are not able to communicate, or they do not exist. To communicate with us, the extra civilizations must be at our stage, or more advanced. It is useless to claim that they exist, but we cannot see them, or that there is a conspiracy to keep us from them. With all the interest from science, telescopes pointing at the sky each night, radio searches, and others, we have not seen reliable evidence. It could change any time, but nothing yet.

Moreover, we need to consider which inventions must evolve, before Stage 5 becomes possible. In a universe that extends to infinity, in all its stages, any invention is possible. However, we must consider the universe that exists as it is now discovered. For instance, humans have evolved over evolution for 4 billion years, but significant evolution was only in the last 2 million years. Any evolution cannot have been much faster than that, allowing for it to happen. Of the inventions, development of the telescope was the most important. Only this can see other worlds up close. Only his can allow fresh images that allow the development of say, the imaginary use of far distances. If development of this on Earth is complicated and disputed, it is hard to account for it in faraway places, where we are not even sure how far life evolved.

Furthermore, biological organs such as vertebrae, feathers, or a four-chamber heart use large changes in genomes to evolve. This requires time, mutation, and ecological and selective pressures to cause the changes. Therefore, most changes occurred in the past with the times and pressures available. On earth, a minor change in DNA resulted in the evolution of not a radically new organ, but also a new device in the universe. Rather, a new learning process, including, ultimately the ability to conceive that a real axis of change could also manifest itself as an imaginary axis, which gave this change a new reality. This resulted in new changes, but especially once telescopes were invented. It was possible with these changes to see a new reality, which had not existed before.

In summary, it was not that just a few alleles altered, and as a result, humans evolved a large brain, to employ speech. Instead, brains have been growing larger, and higher animals have increased intelligence, for many reasons over the history of life. One reason is that when life was simple, genomes could afford to alter a large amount to adapt, because simple life can replicate in huge numbers, so fitness rewards were high. However, as life evolved in complexity, there was intense struggle to reduce the amount of alteration needed to adapt to conditions. Versatile behavior in animals allows rapid adaptation for low genetic cost. Large brains with a high ratio of learning to reflex have furthered the trend. Looking over the universe so far, it is hard to see how peculiarities of life on Earth would be repeated for the universe elsewhere as it exists.

In evolution then, Stage 5 is hard.

There are peculiarities to Earth, such as a large moon, a slow advance of oxygen, a single sun, or shielding by Jupiter. The other problem is that even if advanced civilizations evolve, it is not automatic that they reach science. On Earth, only the western civilization reached modern science, by say, inventing the telescope. Our civilization might also have stopped at Rome, or an earlier stage. If we look around the world, other cultures could stop at any time. Maybe other planets are covered in clouds, or everything seems fine already. Maybe, slave or free, there is no other motive to explore further. Even on Earth, many people are still tied to religion, or do not care. One New Guinea tribe thought that the world was completed by the surrounding mountains, until a plane arrived and landed on the lake. Without proof of another civilization, so far yet, we do not know for sure.

I will leave this speculation at this point. Until we know more, we just cannot tell how life will go.

## *4.2 Is There a God?*

Is there a God, or some force without an explanation, anywhere in the universe? This is a genuine question.

I can point out a force in the universe, but think it is not God, so there must be another explanation. Yet is there a force, we cannot explain at all? I am not myself specifically against God, but we must apply to evolution, as any science, the splitting of facts from the theory. There is no fact that I am aware that can only be explained by God, nor is there a theory that needs God to explain it. From this, fact and theory, it does not justify God. It does not say there is no reason for God on moral grounds, or any other argument why the universe exists.

The problem with this argument against God, though, is that the facts and the theory are never perfect. Philosophers themselves admitted you could never have a perfect set of facts. There must be some aspect, which the facts do not cover. A similar problem was in the theory. If the theory was checked, mathematically, and was complete, yet there could never be a full set of explanations. Somewhere, it has been proven, there had to have incompleteness.

Remember too, the facts and the theory are not the same, and have different methods of proving. If you lined up all the facts and all the theory, it would not prove there was a God, but it might not disprove him either. People who understand these issues must be careful. You can employ facts, or theory, to study each case, but from many studies, there cannot be an absolute conclusion about why the universe exists.

Right now, for instance, even more important than God, the universe is filled with dark matter and dark energy. Oddly, philosophers debated for thousands of years, how we could be sure of things we could not see. Yet it is there. Only 5% of the universe interacts with normal matter. Part of it, the dark energy, is 68% of energy, and this has to do with the expansion of the universe. However, of matter alone, 5% is ordinary, but 27% is dark. If a person promised that they would now explain God, that is fine, but would they explain dark matter and dark energy. Without solving this, I do not see how any explanation can be complete.

However, a person arguing for God this way can be subtle.

In a scientific thesis, I cannot, no one should, accept arguments that are not based on fact or a theory. People can believe anything and argue for flying saucers, or devils from the deep, but assume that the affects here must be scientific, as fact or theory. Even within this, a person could argue how God moved in mysterious ways. The mystery of dark matter, say, might soon be solved, yet there are other factors, we do not know. We might solve dark matter, but not be sure of advanced civilizations. We might wonder about other universes, or if the time is for only this now, or a previous time. Remember, we will never have all the facts, and the theories too are never complete. Even within science, we cannot know everything that is possible.

Therefore, an advocate could argue, we must still have faith in God. Now, I am personally not convinced of a God, but the argument against this is of progressive improvement in science. People can argue that there are things we do not know, but these will be improved, so what is not known now, will be overcome. The source of dark matter, other advanced civilizations, or the causes of evolution, will eventually be found solutions. If we had stuck with God our knowledge would not advance, but by using science, advances are huge, and growing. These arguments too can go forever, so now let me introduce another argument, one that philosophers should have picked up, after the facts and theory. It is that even after the facts and theory are formulated, we still must make a choice over which idea to follow.

Now today, it might seem obvious. You gather all the facts, you test all the theories, and then you make a choice. It might seem obvious, but also look at evolution. After physics, this is perhaps the next most important science, yet people, experts, had a relationship of fact to theory incorrect. Please look in a textbook, any book. It will not state the relationship of fact to theory. It will not explain natural selection as mathematics, and there is no mention of choice. You can watch TV, maybe it will end with a debate of two different theories of why the dinosaurs died, but there is not always a choice in textbooks. To state that philosophers worked out the difference between fact and theory, but evolutionists have this mixed on points, what chance is there that philosophers concluded that facts and theory must end in a choice. If people could not work even that, how is the broader choice applied as action.

Instead, I will propose that fact and theory always end in a choice. Maybe a simple choice, to accept fact and theory as correlated. Maybe though, the choice is harder, such as two different theories, but not enough facts to choose one or the other. Maybe there are too many facts, but no unifying theory to explain them. I propose that this procedure of fact and theory, resulting in choice, applies to science.

Finally, this procedure of fact and theory resulting in choice could apply to the influence of God. Again, I am not aware of any facts or theory that is competently researched, to prove a case for God. However, if just the question is his existence, there is no fact or theory either to disprove it. This is where advocates of God can be subtle. No matter the question, we end up where we face some choice. Even in physics, this applies. The set of facts is never fully complete, and the set of theories always must include incompleteness. You must compare the facts and theory for a choice. Even in the model, there is one explanation for dark matter, but alternatives with a different idea. If this is physics, what happens in fields outside? If you turn to evolution, what is the source of molecular selection to start it? What is the gain for sex? There are theories, but people must compare the facts to the theory, to make a choice.

Now, outside of science, over the world, many people still believe in God. Perhaps others, now and through history, do not believe in the God in the Bible, the Koran, or the Hindu scriptures, but there are questions of which there are no easy answers. Even in physics, people get depressed, one has committed suicide, but pressures also apply in a world without scientific training. Many people today are struggling to survive. My view is that the world is already overpopulated, but you cannot ignore such issues, to claim that there is simply no God. It is not as if the world's problems would all be solved by that expedient. In science, the first task is to separate facts from the theory. If some experts cannot get that right, it weakens the debate about God.

On the other hand, I cannot accept the argument, on facts or theory, which repeats the arguments of ancient writers. I would reach that in any age on the evidence that appears. If you look across religions, there were so many variations of goodness and creation, not one of them could be correct, in an age before science. Eastern religions include the yin-yang of creation, for instance, but western stories ignore it altogether. In the Bible, it is generally Trinitarian, to maintain that God, Jesus, and the Holy Ghost are one person, which is a miracle. This simply will not work in modern scholarship. The story should now be Unitarian at a minimum that at least Jesus was not the son of God, or the story makes no sense in how events happened. I could offer similar the criticism of the Koran, or the Hindu scriptures. Then we must look at the Greek scriptures of how it was different again. There is no point debating each religion, when the stories were from before science.

Rather, we should consider here only the existential argument, that if there is a force we cannot explain, maybe it could be God, or it might be some other argument. I feel that even this, in many ways becomes a choice. This too is often tied to the argument that these views appeal to people who do not have much, so they look for some alternative. The argument that in the end you will go to heaven does not look that appealing anyway. There is good and bad on Earth, or while the bad can be terrible, the good can be more fun than heaven. There could also be reasons for the emotions that we feel. Throughout history, I do not feel that many people have followed the teachings, or in a group, some are more inclined to follow than others are.

In the age of science as well, certain questions can be answered.

We could solve the issue of dark matter. We cold discover evidence of life on other planets. We can solve questions of evolution. Apart from that, for the fortunate, life can be good with science. It almost seems that the more science that we have, the better life becomes. This is part of a general problem of how to make life better. Unfortunately, though, many people still live, even today, in great misfortune. We must find a solution to all of humanities problems.

Let me now offer an opinion, based on the modern debate.

It can be argued, justified or not, that the more oppressive a society in modern terms, the more it tends to offer an opportunity to women to live in a society that some of them might resent. We can see this in modern times. Modern society tells women, which is commendable, that they must be independent, that they should get a good job, that they must fulfil their careers, and so on. However, there is a discrepancy, based on a society that on one hand encourages people, and another one whose premise is "Fear of God". The problem for me is that there is no "Fear of God" that exists in a rational sense. People are told that they must fear the punishment that comes from other members of society. (God does not go around punishing people for disobeying him.) Rather, society makes its rules, and the people who do not obey are punished. Anyone today can experience this. Islamic society is the most extreme, but in any society people who do not follow the rules by an alleged God can be punished.

Another case of this, in societies, is the rule that all women should raise themselves to a higher level, in Asian society, but also in the world. The problem is that men in general do not want to marry them so early. If you are a male with a good salary, there are women available, but often for a younger female to male ratio. Again, there does not seem a setup for this. People can accept a lessor role for women, but not in terms of who or when we marry. In societies that remain traditional, women can marry, and male can try to be loyal, but only to raise a family where the woman does not step outside of traditional roles. Males generally might expect their own female progeny to do well, but he might not expect that in a society which is not bound to traditional norms.

A further argument is how science explains effects that are changing. There was time before when all of science seemed a mystery, but that is changing in the computer age. In evolution, my topic, a mystery was that many effects of science applied across disciplines. However, the effect that carries on is the evolution of sex. My theory says that this can be solved if evolutionists account for the physics of molecular selection. This goes back to the start of the theory. However, if science can reconcile the need of genes to select with an adaptation to how genes will propagate since early life the discrepancies could be overcome. In evolution, this will contrast against the large-scale theories of the universe in physics and in cosmology. In these latter subjects, there are no simple answers. If a person thinks that there are answers, let him explain dark matter or dark energy. The modern debate to me has already gone beyond a thesis that simpler minds can even understand.

By contrast, if we assume that evolution has a scientific explanation, we can assume that eventually the answers will be known. Again, this contrast to physics and cosmology. On the scale of things, we cannot claim to answer the deepest mysteries of existence, if we cannot even frame the questions to ask. If God created the universe, how does he account for effects that no one can even question why they exist. Again, the example, why do modern studies reveal that only 5% of the universe is in terms we would naturally expect. Why is 27% in dark matter and 68% is in dark energy, with no account of why these arise.

Moreover, one can see on the internet questions to which there are no easy answers. It is not just Dark Matter and Dark Energy, but why there is no ratio of relativity to quantum theory years after their discovery, or many other puzzles. An advocate can claim that all answers lie with God, but which God, especially if the advocate cannot in himself answer the questions. I have seen on TV a critic complaining "Well, who is she to question the wisdom of God?" Well, who is he to ask such questions, if he cannot even understand the context in which they were asked? This rests on a presumption that so-called "religious police" have the power, not by God, but their own belief that their law can compel other humans to follow them. If that power is removed, ultimately, all their other power will disappear with that one.

This then is my take on is there a God, but defined I hope in rational terms, as the debate affects humans. I do not wish to seem specifically pro-humanist, but for a free and fair debate, these issues will arise. Today especially, the debate has move beyond how it might have appeared to Newton, or even Einstein. One difference is how the debate if framed. For Newton or Einstein, each was limited to his time be the lack of computers, but today, everyone has computers. If say, we moved to Steven Hawking, he stated against God for the universe to work. If anything, the debate on evolution has been thrown off by the debate over sex. My view, I have stated it here, if we use a model of the universe to prove how the start of sex required an imaginary component of gene spread, the problem of sex can be resolved. If we can resolve this, then solve other debates, over say, evolution, then the debate on the universe shifts to its traditional systems, physics and cosmology. Again, in these systems, advocates of God cannot compete. Or, they can only compete by traditional forms, of advocating control over people's lives by use of so-called religious police, or other, purely human forms of control.

This view then, is my choice, my opinion, about how to present the debate about God, in evolution.

My priority, in science, is to correctly define the relationship of fact to theory, but then to remind people that after these are debated, we must make a choice. If people can achieve this in evolution, the other problems can be solved. If I were challenged to debate evolution, the first questions would be to separate the fact from the theory. Oddly, you might think this is only against the Creationist, but in the crisis now, it must be against experts who refuse to accept molecular selection as an equation. It can be extended to redrawing DNA trees or the advantage of sex. I suggest that many controversies in evolution, are not just creationism, but because molecular selection is still not solved. These consequences, and I have only examined evolution. I do not think that this is general. The debates in physics seem fine to how these are presented, with the different theories, and the choices to be made.

For God himself, I do not have an opinion I can offer, for one way or the other. In the teaching of science, we must stick to facts and theory, but I am not aware of any that justify God. This should be absolute. People can turn to God for other reasons, outside of science, yet to me as well, it ultimately comes to choose. To me right now, over the world, the choices that humanity faces are terrible.

Unless we can improve those factors over the world, God will be in how people act.

## *4.3 Recapitulation*

"For I am well aware that scarcely a single point is discussed in this volume on which facts cannot be adduced, often apparently leading to conclusions directly opposite to those at which I have arrived." **Darwin**

Is the hypothesis in this book valid?

The issue is that life evolved on two axes. There is Darwinian selection, where genes compete to be fit. Then there is an older, pre-Darwinian type of selection, where a gene that can hold its sequence longer than rivals will spread. This contention is not new. People know that genes must compete be fit, but they also know that before cellular selection began, genes also competed to see which ones can hold its sequence. Selection began at the start of life, so this idea is not new.

Instead, the issue is the letter 'j'. If we study evolution, it will only work if part of it is imaginary. Evolution must rise and fall. It must be confirmed, but it seems that molecular selection, which was first, leads fitness to rise. However, Darwinian selection results in the fall of fitness. If it began about 4 billion years ago, the first selection along an imaginary axis caused a rise. When Darwinism joined, this resulted in a fall. We can see this is early genes. If we assume that early selection resulted in a rising logistic curve, the part of selection before Darwinism split genes into competing elements. This now requires a 'j' to explain it, and to split the axes of spread into real and imaginary. If natural selection was from early life, the split applied then, and facts reveal this. From this very early life genes evolved along the two axes.

This way, molecular selection, using a simpler principle was first. In molecular selection, the gene must hold its sequence longer, to evolve by natural selection. Using cellular selection, the gene gains in conservation the more it mutates, but using molecular selection, the gene gains the more it holds its sequence over times. The molecular gain is across all species, but the cellular result is per species. Moreover, if you studied the universe, you would see the problem. If you studied electrical engineering you would see it. Cellular selection is along a real axis, but the molecular selection is along an imaginary axis.

If you use this approach, then evolution will work. It will not only work, but paradoxes, such as for first selection, the modern *neutral* rate, and the evolution of sex will be solved. They can be solved when the axes are divided into cellular and molecular selection.

If anything, there is a simple proof of this new approach, by carefully comparing the facts of how life evolved. Life seemed to begin about 3.7 to 3.5 billion years ago, from cellular selection. Except, life, or prelife, began earlier, perhaps 4.1 to 3.8 billion years ago, by molecular selection. People, experts, know this happened, but there is no model to combine molecular and cellular selection as one. Instead, devices are used, such as life did not evolve until LUCA, or that archaea, bacteria and eukarya began to evolve at once. Experts know that it cannot be true. Rather, because molecular and cellular selection cannot seem to add, this excuse seems good enough. This why we must separate the facts from the theory. Solving this has not become a small problem, but it is huge.

Now, someone can claim, of course, these effects do add, and it needs a slight adjustment and it is solved. However, if a gene gains by 100% in one type of selection, it cannot combine with an earlier type, using an older gain. Experts try to overcome it, by claiming that whatever occurred before would cease, to be replaced by Darwinian selection. Yet if pre-Darwinian selection were for many branches, to absorb heat to gain conservation, it cannot be replaced by Darwinian selection, radiating heat, per species. For the selections to combine requires a mathematical change, but is not there. When I say that this claim is not included, anyone can check in any thesis. The two axes of selection are not used.

Again, as proof, I offer the evolution of sex. Evolutionists write about it knowing it is not solved, but think that it is. Except that, evolution needs two changes. One is to admit that genes increase along two axes. The gain along one axis, cannot explain the dip in evolution, no matter how it is tried. The next is to see that the mutation of mitoses into meiosis is a key to sex. This is a major mutation, and no other factor results in sex. Sex still came a billion years after the evolution of eukarya, which in turn came after an archaeon first ingested the bacterium. Another problem is that for a billion years, there was no diversity in early eukarya. However, once sex evolved, there was wide diversity into many new clades.

Another paradox is the *neutral* theory of molecular evolution.

Again, this has changed, from 'non-Darwinian', and then to *neutral*, but the theory is incorrect. Life cannot increase in gene gain, in prelife, and suddenly, in the modern area, go in an opposite direction. This paradox is less well known, so it is not clear. Rather, the theory is incorrect. There is a *neutral* theory, such that around the neutral point, genes increase the neutral mutations. Here, the theory of molecular selection does apply, but in the new theory this is along a different axis. Apart from this, the theory can solve paradoxes such as the evolution of sex.

Even then, it is strange to challenge evolution on major points of their theory. However, molecular gain is not there. The mathematical model is not there, but even a summary of how molecular selection worked, from prelife, through the transitions since, is not there either. You cannot start, at prelife, without a model, and then carry it through the transitions since, to modern life. This then ends up with a theory that works in a seeming reverse sense, to how evolution now functions.

Suppose though, people might accept that there are paradoxes, and wonder if my theory should be challenged. No other solutions for the paradoxes work. How does a theory of evolution competing along two pathways of selection solve these problems? The point is that if life evolves, it must be gaining in frequency, and wide gene distribution must reflect this. If molecular evolution is *neutral*, gene frequency will stay the same, or in non-Darwinian results it will go down. There have been several major transitions since.

Instead, genes did gain frequency, in two steps.

One was from genes increasing frequency by doubling, such as RNA into DNA, or the single chromosome doubling for higher life. The other was for genes to compete directly for selection, by conserving sequence more than rivals do, to spread by gain of conservation. If an evolutionist has been taught that genes can only gain by mutation, it seems incredible for genes at the same time to gain by conservation. Yet facts support that they do. In eukarya, for example, genes can be rated for how conserved, starting with ubiquitin, then H4, H3, down a scale. If we measure across genes how widely distributed, they are the most conserved genes are the most widely distributed. At the other end, are virus genes, which are highly mutable, but for this there is less distribution, also across life.

The other effect is that from prelife genes competed to gain frequency by conserving sequence, which occurred first. Natural selection began before cellular life, or, if life is from genes mutating, before that, genes first competed to conserve sequence. If anything, from 4.1 to 3.8 billion years, early life competed to first conserve sequence. This also meets the requirements of thermodynamic direction. Change of a sequence, such as conserving sequence, will take minimal energy, and can happen anytime. By contrast, change such as first life, or higher life, takes a mutation to grab energy. An increase in conservation was first, with a thermodynamic advantage, absorbing rather than radiating heat. As the body of life started to mutate, changes that made the new effects would start to take effect. This is the only affect to account for this change as life evolved. Once we accept that genes gain on huge scales by conserving their sequence, and that in the scale of life, conserving sequence was first, we see a different pattern of how genes gain.

The standard theory teaches that genes gain by mutating, but it is not true, in genes that are neutral, will not change, or genes that are non-Darwinian, will lose frequency. In a broader sense though, successful mutations still spread, but they cannot be the widest. All genes in higher life contain ubiquitin or histones 100% spread across the domain, or hox genes 100% spread in fungi and animals, and so on. Subsequent spread of genes must be measured against this. This also explains why there is so much duplication of the genes in higher life. If a gene mutates, even successfully, it loses frequency across life. It is better for an established gene to release a copy, which can alter into a new species, than for the original gene needing to mutate.

Current models can only explain gene spread in one species, resulting from mutation. However, a full model must show that highly conserved genes are laid first, and only from this step, do various changes such as from mutation, the broader segment of life alters. Especially, for a major transition, such as into first life or higher life, highly conserved genes set the transition first. Only after this, did effects such as mutation, solidify the model as sustainable. Sex itself evolved only after the chromosome had doubled. Sex was an adaptation to the chromosome doubled, and the key was the adding of meiosis onto mitosis. We cannot define sex without this. From this step, all the other problems of sex resulted from the changes that formed this. Exercises such as the two-fold cost and other effects, cannot resolve this problem.

Now, let me recap about thermodynamic direction.

To repeat, the equations of gene frequency, cellular or molecular, are reversible, and are not about thermodynamics. In my equations, though, there is an 'angle' between the imaginary and real component. I suggest that this is positive, so that the molecular selection leads the cellular by $90^{o}$. I propose that this shows how life evolves from nonlife. I further propose that for molecular evolution, thermodynamic direction was with a temperature gradient, from hotter surroundings into a lower vesicle. It occurred until the cell formed, to allow a gain by mutation. Until these were complete, the cells could not migrate into a cooler environment, to prove the gain that way.

With this new, combined model, the paradoxes can be solved.

There cannot be a major transition, where gain falls due to mutation, without an increased gain along the molecular pathway. If gain falls, such as the chromosome joining, it goes up in a greater instance, such as enhanced spread over life. Similarly, if the chromosome falls to allow sex, the result goes up, such as for the double chromosome of life. Here, the *neutral* theory is correct, because it gives a $\varepsilon_k = 1$ (approximately). This places genes on the back curve, so they can ascend for gains by sex. A theory of how genes gain in one species, cannot trace the history of life, or cannot solve the other paradoxes.

We can make this objection even stronger.

If scientists reveal that my theory does not work, so there are problems explaining first life, sex, or nonlife into life, we can accept that as the role of science. However, if scientists still claim, based on the old theory of gain only by mutation, that they have already solved this, the claim is not correct. One cannot start from gain for first life, already too high to explain by cellular selection alone. Yet, without explaining the chromosome, nonlife into life, or how life evolved across the transitions, to claim that sex is the only problem left, and this has been solved. One wonders if an obsession with disproving "creationism" has blinded some evolutionists to the role of science. This to solve all the facts before them.

For these reasons, and others, I am convinced that the first molecules competed to conserve sequence. I am also certain that the competition did not cease early on, but has continued down the eras, as the primary molecular competition. I cite many arguments; that genes evolved in ways where conservation increased more than mutation; that conserved genes became widest distributed; that horizontal exchange continues whenever genes cannot obtain high conservation; that the chromosomes enable this competition; and that the paradoxes of the transitions cannot be resolved without considering this. Of the force of these arguments taken together, there is little doubt.

Having developed the model in outline, I can only partially develop this into specific solutions. I have not developed detail solutions, say, for gene doubling in higher life, nor do I solve for thermodynamic direction across life. Instead, I have used the remainder of the book to give a descriptive model of how life evolved, from insights that the theory of gene distribution has given.

In early models, it seemed that a traditional view of fitness must fall, as organisms adjusted, and it is if we use the *neutral* or non-Darwinian model. However, the new model shows that fitness will rise along the axis of gene conservation. Rather, it is too complicated to work out the gains by fitness alone, so we must look instead at many factors, causing the diversity of organisms. Still, it is strange that species evolving complexity seems to be against a loss of fitness. Moreover, if species make a leap of complexity, the changes for this must be huge, so what forces could drive such large losses?

When an open range or existing environment becomes *saturated* with fit types, less vigorous types are pushed into peripheral or less enviable places to survive. Insulated against vigorous competition for long periods, there is an opportunity to experiment, or endure penalties not viable in an open range. Populations forced to evolve in restricted environments enhance properties such as complexity and evolvability. Even so, species do not 'evolve' to become more complex. Individuals attempt to stay simple because it offers the fittest form of life. However, this also sets a 'barrier-pressure-release' for large-scale change, when an environment becomes *saturated* with fit types.

The final section discusses human evolution and behavior.

This was a starting point for investigating gene distribution, because if genes are selfish, it does not explain moral or self-sacrificing behavior in humans. One must also explain why from processes of evolution, or genes for great apes, humans evolved to be so different. The uniqueness of the human form also led to conjectures about the pressures across life driving not just human evolution, but all of life to seek radical ways to adapt. Though the chapters go from pre-biotic life to higher behavior, this investigation began for me in the reverse direction. A force of evolution drove at least one species to evolve along a unique fitness pathway. Still, a digression on human uniqueness is not to infer anything from the main argument. I have a view on this, but nothing in human evolution proves the model of how genes distribute.

If research proves the foregoing thesis valid, what will be the effect on all evolutionary theory?

This is a bold question, because more research must be done. If genes distribute so that the most conserved genes become the most distributed, it is a selective force. The book interprets how it operates. Even if research shows that genes compete a different way, the pattern of scale-molecular distribution will be the key to the large-scale model. Beyond correlating data and verifying the model, we hope that there will be a new large-scale model, of how life evolves. This model should be compatible with existing gene-frequency methods, but also go beyond the paradoxes inherent to that approach. It should be able to resolve the problem of how steady changes at one level produce a varied pattern of how life evolves.

Here, the problem is to show how natural selection produces a theory of selection at one level, giving a higher gain at a new level. We must be careful. I mentioned, if a gene can reproduce its copy, for each replication, the gene will not merely spread, but the gene will reproduce the logistic curve, the essence of selection. For me, I did not study results for evolution along an imaginary axis, but guessed that this must occur. To evolutionists, who studied these effects, you cannot leave half of evolution just blank, as though molecular selection never did occur. There is an effect from early selection, which must have carried on. Once this is solved, then at last, evolution will be complete.

Finally, let me comment on the *neutral* or non-Darwinian theory, as this is also a part of existing theory.

This theory is not correct, if you apply the test of cellular selection to the changes of molecular theory. Yet, once you allow that, there are two pathways of natural selection, the theory works. The previous view has overlooked, in a major way, how genes compete for molecular gain over life, by conserving sequence as they evolve. I cannot see how the existing theory can go on, without a change to allow for molecular selection along the wide pathways mentioned. People love to claim how this works, but it does not fit with how the rest of life evolves. Along a new axis though, molecular selection now works as predicted.

Turning to the paradoxes, none of them relating the gain of frequency by molecular selection has been solved. It is useless, when evolutionists write that these are "almost" solved, when none of them can be. It is useless as well, if the one pathway model cannot solve why the chromosome is a gain, at the start of life, then to jump to solve for sex, far more complicated. Sex too involves many features, such as the longer chromosome for higher life, loss of horizontal transfer, and many effects. This can only be solved by the two-pathway model. It must also account for how the mutation of mitosis into meiosis allowed this. It is strange to claim that it is solved, and not to ask further, when people trying to solve it do not even consider the two-pathway model.

It is even worse. In front of the world, I have challenged evolutionists to prove how certain paradoxes are solved, but I provide a suggestion. All anyone needs to do is prove that my suggestions are not necessary, and that these have been solved, and the claim is defeated. Using a frequency approach to small-scale, to model large-scale observations, is long overdue in evolution. The model here is a way forward.

It has shown, not just in modern selection, but also in a molecular struggle in the prelife of conservation, why billions of years later "endless forms most beautiful" have also been, and will continue to be, evolved.

## Appendices

# *Appendix I – Brief Technical Background*

### Using Large Numbers

This book uses the scientific notation for large numbers. Some rules are;

1. *Any number is its value in units multiplied by the power of 10.* A million has six zeros, so we write it as $10^6$, one with six zeros. Three and a half million people is 3 ½ times one million, or $3.5 \times 10^6$ people.
2. *When numbers are multiplied, powers are added.* Three and a half million people with thirty thousand genes each is a total of; $3.5 \times 10^6$ times $30 \times 10^3 = (3.5 \times 30) \times 10^{(6+3)} = 105 \times 10^9$ This in turn = $1.05 \times 10^2 \times 10^9 = 1.05 \times 10^{11}$ total genes.
3. *When numbers are divided, powers are subtracted.* If a fragment is $10^2$ bp long and it mutates at $10^{-4}$ over all regions, then in $10^2/10^{-4}$ or $10^2 \times 10^{-(-4)} = 10^6$ reproductions, it will obliterate its sequence.
4. *When numbers are inverted, the power changes sign.* A gene mutates once every two million generations. Its mutation rate is; $1/(2{,}000{,}000) = ½ \times 1/10^6 = 0.5 \times 10^{-6} = 5.0 \times 10^{-5}$ mutations per generation.

Remember;

a) A million = $10^6$ and a (USA) billion = $10^9$.
b) A trillion = $10^{12}$ (a thousand billion)
c) Gene lengths are in bases (or base pairs, bp). A Mega base is $10^6$ bp and the human genome is about 3,000 Mega base or $3 \times 10^9$ bp.

### The Second Law

The most misunderstood law of physics concerns three things;

a) The First Law states that we cannot get energy for nothing.
b) The Second law states that even if there is enough energy, some of it will be lost due to inefficiency and dissipation.
c) The principle of Thermodynamic Equilibrium (TE, but part of the Second law) states that systems will tend towards disorder.

Total available energy in the universe is a function of its absolute temperature, but this falls over time. Today the universe is about $2.7^0$K, but when Earth formed, it was about $3.3^0$K. The universe has consumed $3.3 - 2.7 = 0.6\ ^0$K of equivalent work over that time. This has gone a small part into organization that allows life. Yet if there was a nuclear war that disorganized life again, the $0.6\ ^0$K would not return to the universe as extra temperature, but would dissipate away. (The mean temperature in the universe will never increase while the galaxies expand, and now it seems likely they will expand forever.)

However, temperature also contains a quality called "order". This means that the temperature difference $T_2 - T_1$ between a hot body and its surroundings is available for useful work. The Second Law states that when $T_2 - T_1$ exists, the statistically likely state is that the difference will diminish over time. (A hot meal in a cold room will cool over time). The state of $T_2 = T_1$ (no temperature difference left) is called Thermodynamic Equilibrium (TE). Life moves against the direction of TE, but it does not violate the Second Law (this is much misunderstood).

**Evolution**

Evolution is accumulation of thermodynamic order over time. Total order in the universe falls with absolute temperature, but local order can accumulate. Initial accumulation is formation of atoms, stars and galaxies. This process continues pre-biotic (prior to life) up to a complexity level of amino acids. Beyond amino acids, complex forms must self-replicate, or self-renew, to maintain order against other Second Law effects. Each step away from TE presents a barrier, which must overcome. From the initial thermal soup of the universe the structure of amino acids is far from TE, so it took billions of years of stellar evolution for complex molecules to form. Order beyond amino acids is not possible until after life evolves. For intelligent beings to build cities, or send a living being into outer space, is a huge step against TE that cannot be overcome by biological evolution alone, but requires social evolution.

**Complex Numbers**

*"Minus times a minus is a plus, for reasons we shall not discuss."*

Well, we can discuss it. $-1$ is a rotation of $180^\circ$ (go the opposite way). Minus times minus is rotate $180^\circ$ twice (come back to $0^\circ$). So, minus times minus is a plus, but it also explains what $\sqrt{-1}$ is. If $180^\circ$ is $-1$, then two rotations of $90^\circ$ give $180^\circ$. So $\sqrt{-1}$ is $90^\circ$, because $\sqrt{-1} \times \sqrt{-1} = -1$.

The values of $\varepsilon_i$ can be listed in relationship to $\mu_i$, but also as an angle;

$\varepsilon_k = \infty$, $\mu_k \approx 0$ ("forever"), $\theta_k = 90^\circ$
$\varepsilon_k = 1$, $\mu_k \approx \bar{\mu}$ ("average"), $\theta_k = 45^\circ$
$\varepsilon_k = 0$, $\mu_k \approx 1$ ("each reproduction"), $\theta_k = 0^\circ$

## *Appendix II –Suggestions for Further Reading*

Note: This is not an exhaustive list of books, but a sample of books in my library. The list does not include general texts, references, or materials from periodicals and the Internet. Nor does the list include well-known, works quoted in the text. Obviously, a great many original works influenced this present book, and any omissions in the following list will be gratefully acknowledged by the author.

Evolution, Monroe W. Strickberger, Jones and Barlet.
Evolution; The Four Billion Year War, Majerus, Amos & Hurst, Longman.
The Origin of Species, by Charles Darwin
The Selfish Gene by Richard Dawkins, Oxford Univ Press.
The Blind Watchmaker : Why the Evidence of Evolution Reveals a Universe Without Design by Richard Dawkins, W W Norton & Co
The Extended Phenotype : The Long Reach of the Gene (Popular Science), by Richard Dawkins, Oxford Univ Press
River Out of Eden : A Darwinian View of Life (Science Masters Series) by Richard Dawkins, Lalla Ward (Illustrator), Basic Books.
Climbing Mount Improbable, by Richard Dawkins, Lalla Ward (Illustrator) W W Norton & Co.
Life : A Natural History of the First Four Billion Years of Life on Earth by Richard Fortey, Knopf
Darwin's Dangerous Idea : Evolution and the Meanings of Life by Daniel Clement Dennett, Touchstone Books,
The Dragons of Eden : Speculations on the Evolution of Human Intelligence, by Carl Sagan Ballantine Books (Mm)
Cosmos by Carl Sagan (Mass Market Paperback - September 1993)
Broca's Brain : Reflections on the Romance of Science by Carl Sagan (Mass Market Paperback - October 1993
Becoming Human: Evolution and Human Uniqueness, Ian Tattersall, Harcourt Brace.
Wonderful Life : The Burgess Shale and the Nature of History by Stephen Jay Gould, W W Norton & Co.
Bully for Brontosaurus : Reflections in Natural History by Stephen Jay Gould, W W Norton & Co.
Ever Since Darwin : Reflections in Natural History by Stephen Jay Gould, W W Norton & Co.
Reinventing Darwin : The Great Evolutionary Debate, by Niles Eldredge, Phoenix.
On Human Nature by Edward Osborne Wilson (Mass Market Paperback - October 1988)

Reading

Population Genetics : A Concise Guide, by John H. Gillespie, Johns Hopkins.
The Causes of Molecular Evolution, by John H. Gillespie, Oxford.
Principles of Population Genetics, by Daniel L. Hartle, Andrew G. Clark, Hardcover
A Brief History of Time by Stephen Hawking, Bantam.
At Home in the Universe: The Search for Laws of Self-Organization and Complexity by Stuart Kauffman Oxford Univ Pr (Trade)
William H. Calvin, The Cerebral Code: Thinking a Thought in the Mosaics of the Mind (MIT Press)
Windows on the Mind : Reflections on the Physical Basis of Consciousness by Erich Harth.
How the Mind Works, Steven Pinker, Penguin
The Emperor's New Mind, Roger Penrose, Oxford.
Not in Our Genes, Rose, Kamin and Lewontin, Pelican
The Ascent of Man, Jacob Bronowski.
From Brains to Consciousness, Steven Rose, Pelican
How the Leopard Changed its Spots, Brian Goodwin, Phoenix
Artificial Life, the Quest for a New Creation, Steven Levy, Pelican
The Red Queen, Sex and the Evolution of Human Nature, Matt Ridley
Shaping Life, John Maynard Smith, Weidenfeld and Nicolson
The Children of Prometheus, Christopher Willis, Perseus Books
The Human Brain, Susan Greenfield, Phoenix.
The Sixth Extinction, Richard Leakey, Phoenix
The Third Chimpanzee, Jared Diamond, Vintage
The Human Brain, Phoenix, Greenfield, Susan,
Evolving Brains, Scientific American, Allman, John
The Cooperative Gene, Free Press, Ridley, Mark
What Evolution Is, Basic Books, Ernst Mayer
The Origins of Life, Cambridge Uni Pr, Freeman J Dyson
Endless Forms Most Beautiful: The New Science of Evo Devo and the Making of the Animal Kingdom; W. W. Norton, Sean B. Carroll
The Making of the Fittest; W. W. Norton, Sean B. Carroll
Darwinian Dynamics; Princeton, Richard E Michod
Life on a Young Planet; Princeton, Andrew H. Knoll
The Vital Question; Profile Books, Nick Lane

# Index

Index

www.ingramcontent.com/pod-product-compliance
Lightning Source LLC
Chambersburg PA
CBHW030013190526
45157CB00016B/2599
*9781090183033*